Islands Down East

Islands Down East

A Visitor's Guide

Text and Photography
by
Charlotte Fardelmann

A Peter Randall Book/Down East Books

To my parents, Ted and Clara Lyman, with whom I first explored Islands Down East aboard Voyageur.

Printed in the United States of America

ISBN: 0-89272-189-8

Library of Congress Card Catalogue Number 84-70016

Maps by Alex Wallach

A Peter Randall Book
Down East Books
Camden, Maine 04843

Contents

Front cover: Young lobstermen, Matinicus.
Rear cover: Visitors leaving Baker Island.
Page one: Carver's Harbor, Vinalhaven.
Page 2–3: Foggy Day off Vinalhaven
Page 4: Monhegan cliffs.
Page 5: Windjammer, Fox Island Thorofare.
Page 6: Near Isle au Haut.

Introduction

AN ISLAND VACATION brings to my mind the image of a long sea journey to an unknown country. Islands Down East suggest the special flavor of granite rock, spruce woods, bell buoys, lobsters, and Down East humor.

Some travelers to these isles seek peace and quiet, a chance to read books and take long walks. Others are looking for adventure, places to go and things to do. This book aims to help the first-time visitor make an intelligent choice of where to go and how to arrange it.

I have focused on 22 islands along the Maine and New Hampshire coast. All are accessible by public ferry or regularly scheduled commercial boat. For yachtsmen using this book, information on moorings, marine repair service, and fuel facilities is included.

Each chapter includes my impressions, a brief history, and a description of inns, restaurants, facilities for yachtsmen, ferry service, things to do and points of interest. The last chapter includes special programs and vacations through which one may reach additional Down East islands. The back of the book also includes a special section with rates for transportation and accommodations.

The islands of Maine and New Hampshire are familiar to me. Although I grew up in Minnesota, far from salt water, I have been summering on, around, and near Maine and New Hampshire islands for over thirty years.

My first introduction to Maine islands came when my parents bought a beautiful sloop, a "New York 32," and began annual summer cruises in New England waters. It was aboard *Voyageur* that I first became enchanted by Maine islands. After marriage, I continued to cruise the Maine coast with my husband and four children.

Eventually we purchased land at Crockett Cove on Vinalhaven and experienced the joys of summering on a Maine island. Grocery shopping was by Boston Whaler through Leadbetter Narrows to Carver's Harbor. We swam in deep freshwater pools that had formed in the island's abandoned granite quarries. We climbed the cliffs behind our house and picked raspberries, blueberries, and blackberries. Mucking around in rubber boots, we explored the sea life of our cove. We lived by sea time in the rhythm of the tides.

Our pace grew slower. On islands, socializing is an important part of every endeavor. A man who came to repair the stove would take time to enjoy a cup of coffee and a piece of pie.

When we couldn't find what we wanted in the island store, we learned to make do, avoiding at all costs that trip to the mainland. An easy dinner was always available by picking the blue-black mussels off slippery wet rocks at low tide from shores untainted by pollution.

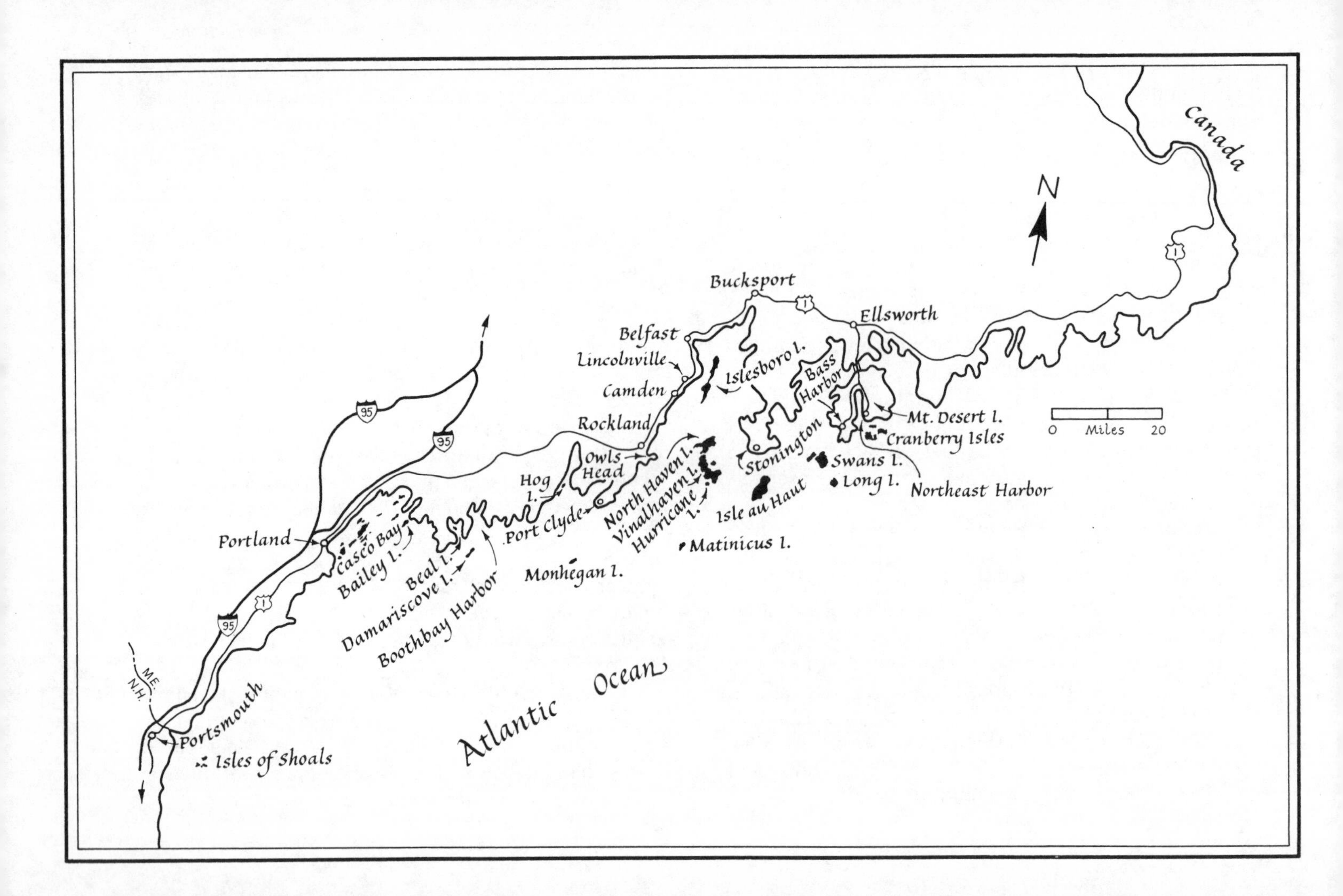
Canada
N
Bucksport
Ellsworth
Belfast
Lincolnville
Camden
Rockland
Islesboro I.
Bass Harbor
Mt. Desert I.
Cranberry Isles
0 Miles 20
Owls Head
Hog I.
North Haven I.
Vinalhaven I.
Hurricane I.
Stonington
Swans I.
Long I.
Northeast Harbor
Isle au Haut
Matinicus I.
Port Clyde
Monhegan I.
Portland
Casco Bay
Bailey I.
Beal I.
Damariscove I.
Boothbay Harbor
95
1
ME
N.H.
Portsmouth
Isles of Shoals
Atlantic Ocean

I particularly recall certain islanders who couched their communications in a wonderful dry, soft humor aimed at delighting and entertaining everyone within listening distance.

The island I visit most frequently now is Star Island, one of the Isles of Shoals. Only ten miles out to sea from my home in Portsmouth, New Hampshire, this wonderful treeless island is populated primarily by gulls and Unitarians, in that order. Star is a wonderful off-shore island steeped in history and spiritual pursuits.

Island people are friendly. On nearly every Maine island, people wave to you from their cars. At first I thought they mistook me for a friend, but now I realize that they treat everyone this way. No one lands on an off-shore Maine island unnoticed. On such an enclosed space, every person, tourist or islander, has some small effect on every other person there.

Sometimes you don't know the effect; sometimes you do. One day, afraid of missing the last boat from Swans Island to Bass Harbor, I pulled my car quickly out of the driveway in front of a car coming rather slowly along the road. I sped down to the ferry terminal only to find I was early. A few cars were waiting in line and the ticket booth was closed. Minutes later, a lady arrived and opened the ticket window. "Pulled out kinda fast, didn't you?" she said with gentle good humor. "Might 'ave run us down."

Weather on these islands is changeable. Temperatures in summer run around 65 to 80 degrees, but they may drop to the 50s on foggy or rainy days. Wool sweaters and rain gear are suggested for island travelers. Usually the most fog comes in June, while August has more sunshine. Many people think the best time is September, with its brilliantly clear north wind days. The weather is a source of continual speculation and provides a good opener for a conversation with local people.

"Do you think the fog will lift?" I might ask. The response might be delivered with a twinkle in the eye: "Eh-ya, always has."

Island communities are far from idyllic. They are real places with real problems and real people. Food and services are expensive, jobs are scarce, and most enterprise involves trips to the mainland.

Most of the islands count on lobstering, fishing, and summer people for income. Tourism is an additional boon to the economy for some places. On the smaller islands, where the economy is uncertain, whole communities hang in the balance when one or two families leave to seek better employment opportunities. When you have fifty year-round residents and eight children, one young family moving off-island can have a huge impact on the community. The school may close for lack of the minimum number of children required by the state. Without a school, other families may have to move too. Old folks dependent upon these young families may have to go with them. When the vacated homes of these families are purchased by outsiders as vacation homes, the year-round community is diminished. The next phase is "summer island," as the winter community becomes a ghost town populated only by a caretaker who checks the cottages for vandalism or storm damage.

The ferry is the link between island and mainland. At ferry time, half the community stands around to see who gets off or on. Events are scheduled around "the 10:45 boat" or the "late afternoon boat." The number of ferry trips depends upon the tickets sold as well as the distance from the mainland. Peaks Island, next door to Portland, has 26 boats a day in summer. Islesboro has nine daily trips. On the other hand, Long Island (off Mount Desert Island) has only two boats a week scheduled on different days, making trips to the mainland difficult for the islanders and day visits by tourists hard to arrange. Ferry service to Matinicus is almost nonexistent now; a visitor must go by plane.

Webster defines ferryboat as "a vessel for conveying passengers, merchandise, etc. across a river or other narrow water." The ferries along the coast of Maine and New Hampshire cover a wide spectrum of design. Some are relatively small: the Isle au Haut and the Cranberry Isles boats carry 50 to 75 passengers; the mail boat to Monhegan could be taken for a fishing boat. Others, such as the Casco Bay Lines double-decker cruise vessels, can take 200 to 300 passengers.

The Maine State Ferry Service provides ferry transportation to five residential island communities: between Rockland and Vinalhaven; Rockland and North Haven; Lincolnville and Islesboro; Bass Harbor and Swans Island; and Bass Harbor and Long Island (Frenchboro). Once a month a trip is scheduled between Rockland and Matinicus.

On every island the biggest worry is fire. An island fire started by a careless smoker or camper is not easily put out. Volunteers fight the fire with minimal equipment, usually one truck. By the time the fireboat arrives from the mainland, it may be too late. Fires may keep burning underground for a long time. Isle au Haut was burned over by blueberry pickers in the summer of 1870 and the fire was not finally extinguished until the winter snow.

If fire is the biggest hazard from tourists, litter is the biggest annoyance. Dropping lunch bags and empty cans is an act of obscenity on these wild and natural shores.

I had expected more culture shock between the mainland and the islands. That might have been true thirty years ago but television has changed things. Islanders watch the same newscasts, the same movies, the same soap operas, and the same talk shows as do people everywhere—if they have electric power. Most of the islands featured in this book have cables to the mainland. Isle au Haut's cable was laid in 1983. Matinicus, too far out to sea for a cable, has a central generator. Out of all the year-round island communities I visited, only Monhegan has no central power. Monhegan islanders buy their own private generators; a very few use photovoltaic cells.

Telephone service goes to nearly every year-round community listed in this book. The only exception is Isle au Haut. In 1983 Monhegan upgraded its poor telephone cable system to a modern microwave (radio-transmitted) system. Frenchboro received phone service for the first time in 1982.

Escalating property taxes are another concern of island residents. As the coastal land boom moves eastward, shorefront island property has grown in value. Rising taxes threaten to drive out the year-round native islanders. On Cliff Island and several others in Casco Bay, residents are taxed as part of the city of Portland although they receive almost no city services.

Attitudes toward tourists vary. I have indicated in this book which islands have the most facilities for day visitors and which islanders would just as soon not have tourists land at all. Visitors are urged to observe rules about fires, camping, pets, and trespassing. From the point of view of the tourist, I suggest the most pressing problem on almost every island is the lack of a public bathroom.

My research of several months on Maine islands had a deep effect on me. The beauty is haunting — the dark silhouette of a spruce tree against the sparkling sea; a line of smooth pink granite ledge between dark green forest and blue water; a tiny tidepool filled with lime-green algae and blue-black mussels; a lobster boat perfectly mirrored in a quiet harbor. I made friends and learned lessons. I hope some of their thinking — and humor — rubbed off on me.

These islands draw me back. I suspect that I am not alone in my yearnings. People fall in love with islands. There is something romantic about them. A New Yorker who summers on Monhegan says she sometimes wakes up, in her city apartment, mentally hearing the gulls cry and feels compelled to pack her bags and head for her island.

When I stand on the granite boulders that tumble out into the open ocean from one of my favorite islands, I listen to the wind sweep through the spruce forest and watch the sea crash, wave after wave, on the ledge below. The sea, the rock, the wind, the spruces — these are things that move my spirit, catch my gut, and brand my memory.

For me a few days on a saltwater island is a chance to get in touch with my own deepest self, that inner island where priorities become clear and creativity is born.

Acknowledgments

I want to thank islanders and others who helped me research or were kind enough to proofread my chapters: John Heiser, Arthur Borror, James Smith, Jane McDermott, Gretchen Hall, Hilda Cushing Dudley, Wayne Selberg, T.T. Rand, Daniel and Kay Carr, Mark Greene, Mary Justice, Edmund Doughty, Johanna E.R. von Tiling, Sarah Burnham, Lexi Krause, Larry Cooper, David and Katy Boegel, Ed and Ann Hubert, Ruth Boynton, Josephine Day, Lynn Drexler, John Hultberg, Ade Mersfelder, Gladys Mitchell, Betsy Burr, Geoffrey Katz, Pat and Arthur Crossman, Thelma Burgess, Almon Ames, Lewis Haskell, Jon Emerson, Greg Marquise, Bill Warren, George Martin, Henry and Margaret Hatch, Shelley Woods, Danny and Tina Lunt, Rebecca Lunt, Vivian Lunt, Marie La Rosee, Alberta and Ted Buswell, Leona and Carroll Lilly, Sonny Sprague, Llewellen Joyce, Norman and Peg Bailey, Karen Fernald, Jane Porter, Ted Spurling, Conley Salyer, Marion Emerson, Polly Gates, and others.

I want to express my appreciation to Peter Randall who encouraged me from the start and kept his good humor throughout the production of this book; to Karin Womer of Down East Books for her editorial suggestions; and to Alex Wallach for his patience in revising the maps.

Everyone enjoys rocking on the wraparound porch of the former Oceanic Hotel, now the main building for religious conferences run by Star Island Corporation.

The Isles of Shoals

ISLANDS DOWN EAST begin with the Isles of Shoals. The only major islands between Boston and Portland, the Isles of Shoals stand a lonely vigil seven miles off the mainland. The Maine-New Hampshire border runs through the middle of this island cluster so that five of the nine islands belong to Maine and four to New Hampshire. Small, bleak in appearance, and relatively barren of vegetation, the Isles of Shoals have played a surprisingly significant role in history.

Before the Pilgrims landed on Plymouth Rock, the Isles of Shoals were busy stopping-places for fishing vessels, harbors at which fish were dried and salted for the European market. Settled year-round by the early 1600s, these islands have supported residents for three hundred and fifty years.

During the nineteenth century, the Isles of Shoals became a popular cultural center for the leading writers and artists of the day. Many were drawn to the Appledore House by the charms of resident poet and author Celia Thaxter. In her book, *Among the Isles of Shoals*, Thaxter describes a first visit to these islands:

"Swept by every wind that blows, and beaten by the bitter brine for unknown ages, well may the Isles of Shoals be barren, bleak, and bare. At first sight nothing can be more rough and inhospitable than they appear. The incessant influences of wind and sun, rain, snow, frost and spray, have so bleached the tops of the rocks, that they look hoary as if with age, though in the summer-time a gracious greenness of vegetation breaks here and there the stern outlines, and softens somewhat their rugged aspect. Yet so forbidding are their shores, it seems scarcely worth while to land upon them, — mere heaps of tumbling granite in the wide and lonely sea. . . ."

Thaxter remarks, "Landing for the first time, the stranger is struck only by the sadness of the place,—the vast loneliness; for there are not even trees to whisper with familiar voices, — nothing but sky and sea and rocks. But the very wildness and desolation reveal a strange beauty to him."

Today, the Isles of Shoals boast two educational institutions: the Shoals Marine Laboratory (SML), jointly operated by Cornell University and University of New Hampshire, and Star Island Conference Center for Unitarians and members of the United Church of Christ.

The *Viking Sun* out of Portsmouth, New Hampshire, provides the ferry service. From the open upper level of this three-deck, 500-passenger cruise ship, there are excellent views of the waterfront. As the boat slides down the Piscataqua River between Kittery, Maine, and Portsmouth, Captain Whittaker gives a running commentary on the tugboat fleet, nuclear submarines at Portsmouth Naval Shipyard, an abandoned Navy prison, several forts, and other points of interest. Lighthouses on either side mark the river mouth as the *Viking Sun* rocks gently into the ground swell of the Atlantic Ocean.

The one-hour boat trip across to the Isles of Shoals can

Isles of Shoals

Star Island

1. Wharf
2. Star Island Conference Center (Old Hotel)
3. Meetinghouse
4. Summer house
5. Betty Moody's Cave
6. Tucke Monument
7. Vaughn Memorial
8. Captain John Smith's Monument

— Road

Appledore Island

1. Boat landing
2. Kiggins Commons (Rest Rooms)
3. Foundations of Appledore Hotel
4. Laighton Cemetery
5. Celia's Garden
6. World War II Tower
7. Rookery Trail
8. Swimming pool

— Road ∘∘∘ Breakwater
---- Trail

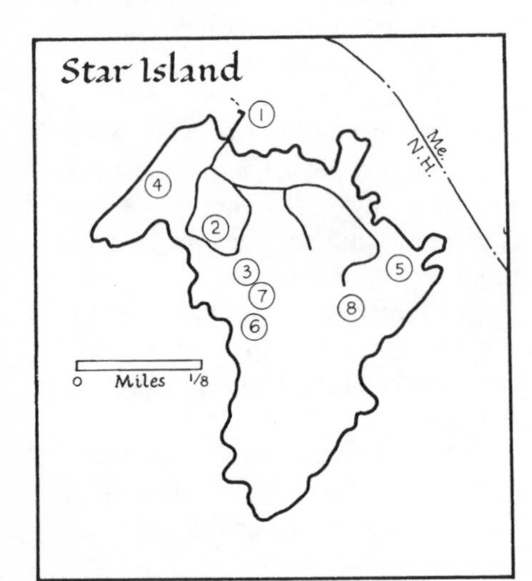

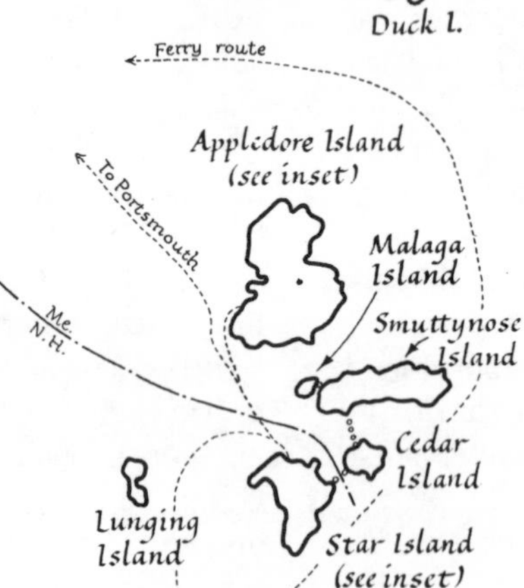

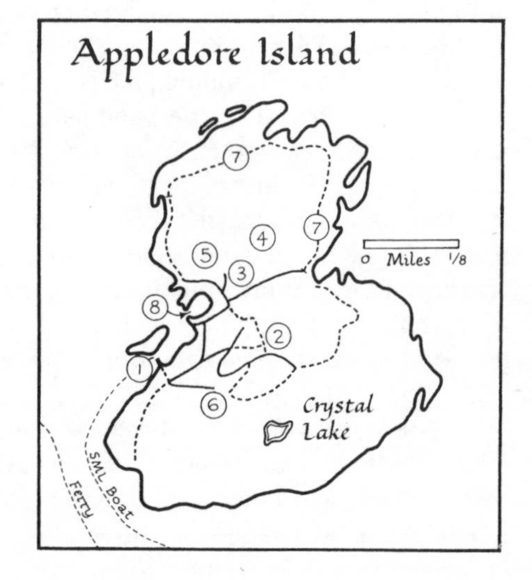

be bumpy on some days, on others so smooth that every buoy is reflected as in a mirror. At some times the islands appear sharp and clear on the horizon, at other times they are hidden behind a woolly blanket of thick white fog.

The Islands

In contrast to many islands that have year-round communities with shops and public roads, the Isles of Shoals have no such towns. The year-round resident population lasted from the early 1600s until the 1870s when the few remaining residents of Gosport sold out to hotel interests. All nine islands are privately owned, except for White Island, which is owned by the United States government. Star and Appledore are open for visits by ferry passengers during the summer season. The vast majority of visitors stop only on Star Island, but people interested in seeing the Shoals Marine Laboratory may make prior arrangements to be ferried to Appledore in SML's small boat. (Contact University of New Hampshire Marine Program in Durham, (603) 862-2994). Both islands are small enough to walk around easily in an hour or two. The largest of the Isles of Shoals, Appledore, is 95 acres, and Star is 40 acres; all nine islands together total only 205 acres.

Star Island

As the *Viking Sun* pulls into Gosport Harbor and docks at Star Island wharf, only "stopover passengers" are allowed to get off the boat. Star Island Corporation, which owns this island, operates busy week-long religious and educational conferences (up to 240 participants), and strictly limits the number of day tourists (see "How to Get There").

To land at Star Island is to be transported back into the nineteenth century. Facing the harbor is a four-story, white clapboard, Victorian hotel, complete with wraparound porch facing the harbor. A little summer house sits on a knoll overlooking the sea, offering a tiny bit of shade on a bright day. Conferees quietly walk the paths, while their children play games on the open meadows. The broad treeless vista has a strange, otherworldly look.

About ten thousand years ago the last glacier to come through this area sheared off the top of an ancient alpine mountain and dropped it unceremoniously onto the nearby ocean floor. What was left became the Isles of Shoals. Scars left by glaciers can be seen on many rocks.

Over the years, the islands have gained a thin layer of soil that seems to produce beautiful, deeply colored wildflowers, lots of berry bushes, and over two hundred species of vascular plants including a number of shrubs and stunted trees. (Visitors are requested not to pick flowers or plants.) Poison ivy is especially lush, but it has been cleared back from paths.

The old hotel has the ambiance of a nineteenth-century island retreat, with its tall ceilings and echoing halls. The college-age young people who staff the hotel are called "pelicans." The staff also includes many "old shoalers" who have come each summer for decades and like to talk about the island. Fred and Virginia McGill have been summering on Star for over fifty years. Virginia McGill's pamphlet, *The Spirit of the Shoals* (available in the bookstore), makes interesting reading material for a quiet half-hour in one of the rocking chairs on the hotel porch.

Star is a wonderful picnic island, with its quiet coves overlooking the harbor, its flat ledges, high cliffs, and lovely wild flower meadows.

Children always love Betty Moody's cave, for they can crawl in one end and out the other. This cave was named for a woman who hid there during an Indian raid in 1724, and who, in an effort to silence her youngest child, accidentally suffocated the infant.

I always take time to meditate in the quiet of the small native-stone meetinghouse, built in 1800 on the highest spot on the island. Before there was a lighthouse on the Isles of Shoals, villagers placed oil lamps in this meetinghouse each night as a beacon to their seafaring menfolk and other sailors. This tradition continues every summer evening as conference participants line up at the foot of the hill holding candle lanterns. The procession winds in silence up to the meetinghouse, where the lanterns are hung from wooden sconces during a brief service. Later, the lanterns are carried silently back down the hill.

Go through the old turnstile that used to mark the edge of the pasture land and follow the trails out toward the back cliffs. This is gull territory. Be careful to stay clear of nests and young chicks during spring and early summer. Gulls vigorously protect their nests by diving at intruders, sometimes defecating on them or even injuring them. A stick held overhead is usually enough to keep the birds from striking. The shore areas beyond the Captain John Smith Monument are best avoided early in the summer.

One of the indentations on the cliffs is called "Miss Underhill's Chair." It was named for a Gosport schoolteacher,

Candlelight services are held each summer evening in Star Island's stone meetinghouse, built in 1800.

Miss Nancy J. Underhill, who was washed away from this favorite outlook by a huge wave that broke over the cliff in 1852. Her body was found a week later at Cape Neddick on the coast of York, Maine. Children should be supervised when out on the cliffs, especially on the exposed shore. Unexpectedly high waves could carry an unsuspecting person into the ocean.

The path winds back through high shrubs to a tall granite obelisk. This monument was erected in memory of the Reverend John Tucke, the minister on Star from 1732 to 1773. Another memorial marks Captain John Smith's visit in 1614. The Celia Thaxter room in the Charles F. Vaughn Memorial is devoted to paintings, writings, and possessions of the Isles of Shoals poet (open on limited basis to day tourists).

Appledore

You can recognize Appledore, the largest of the Isles of Shoals, from a distance because of the tall World War II era tower that was used for submarine watch. This island was once named Hog Island, presumably because its rounded appearance resembled the humped back of a hog wallowing in the brine. Thomas Laighton changed the name in 1847 when he planned to build a hotel. (Who would want to vacation on Hog Island?) Almost all of Appledore is owned by the Star Island Corporation and leased to the Shoals Marine Laboratory. Visitors are welcome during daylight hours of operating season, late May to mid-September.

To be on Appledore for even ten minutes is to understand that the real owners of the island are the gulls. Calling and screeching, they provide a continual background noise everywhere on the island. Dr. Arthur Borror, Associate Director of Shoals Marine Laboratory, has made a special study of the birds of these islands. Borror claims there are close to six thousand pairs of nesting gulls on the Isles of Shoals (3,800 pairs of herring gulls and 1,800 pairs of great black backed gulls), one of the largest concentrations of gulls in the eastern United States. About two thousand of these gulls nest on Appledore. When you compare this number to the 60 or so humans who live there, you can understand why people carry sticks above their heads while walking the trails.

Despite these minor distractions, Appledore is a bird watcher's paradise. In addition to gulls, there are snowy egrets, black-crowned night herons, glossy ibis, and little blue herons, to say nothing of swallows, warblers, and all the birds that migrate through. Over 200 species of birds have been spotted on the Isles of Shoals since 1966. One of the projects of the Shoals Marine Laboratory is banding of migratory birds, a project recently expanded to the gull population as well. Appledore's shores are considered Critical Natural Areas by the State of Maine, and Appledore Island is a registered National Historic Site.

Director John Heiser, a tall, genial, red-bearded professor from Cornell University, says, "Most day visitors try to make visits too short." Heiser recommends coming on the early boat at 7:30 a.m. and returning on the evening trip, taking the time to walk the trails, sit on a rock, join in with some classes or field trips, and talk with students.

We walked the rookery trail, following the red blazes as we dodged poison ivy from below and gulls from above, then

White Island lighthouse is attached by enclosed walkway to the keeper's cottage.

cooled off with a swim in the tidal pool (warmer than the ocean) before returning home.

Other Islands

Duck Island, north of Appledore, is so low in elevation (ten feet above high water) that high seas sometimes wash right over it. These eleven acres are home to seals and 1100 pairs of nesting double-crested cormorants. The pond in the center of the island is a favorite migration stop for ducks. This island is managed by the laboratory as a wildlife refuge.

At the entrance to Gosport Harbor is Lunging Island (seven acres). Once called Londoners Island, this was the first fishing base for the London Company, thought to have been established around 1615, and is thus the first "London" in the New World.

The picturesque lighthouse on White Island is a favorite with photographers. The *Viking Sun* passes close to the island on its return trip to Portsmouth. White Island and Seavey Island (13 acres total) are joined at low water. These islands are owned by the Federal government and are manned year-round by the Coast Guard.

Gosport Harbor is formed by breakwaters that join Star Island to Cedar, Smuttynose (27 acres), and tiny Malaga. Lobster boats operate out of Cedar Island during spring, summer, and fall.

Smuttynose is named for black rocks on the western end

of the island. The Malaga-Smuttynose breakwater was built about 1800 by Samuel Haley, who paid for the work by selling silver bars he found in the rocks. The silver bars were thought to be pirate treasure left by one of the pirates who landed on the Isles of Shoals.

Smuttynose is probably best known as the scene of a famous murder in 1873, when itinerant fisherman Louis Wagner rowed a dory twelve miles from Portsmouth to the island one moonlight night, brutally murdered two women with an axe, and rowed back to Portsmouth. Only six people, three men and three women, resided on the island, and the men were ashore in Portsmouth at the time of the murder. One woman escaped, shivering all night behind a rock with only her little dog to keep her warm. She lived to tell the tale, and Louis Wagner was caught, tried, convicted, and hanged at the last execution in Maine before capital punishment was abolished.

History

The Isles of Shoals made a favorite stopping place for fishermen in the early 1600s. The name Isles of Shoals refers to the plentiful shoals (or schools) of fish. Martin Pring mentioned the Isles of Shoals in reporting his voyage to the Piscataqua River in 1603, and Captain John Smith gave a more detailed report after his explorations in a small boat in 1614, under the auspices of the London Company.

The Isles of Shoals became the fishing capital of America between the years 1615 and 1620, with a trading post on Londoners Island. One account states that Miles Standish came to the Isles of Shoals to buy provisions for the Pilgrims at Plymouth in 1623.

This was a fishing station during the early years before permanent residents settled. As many as six ships anchored in Gosport Harbor while the fishermen filleted, salted, and sun-dried the fish before they were packed for the European market. A particularly good type of preserved cod was dunfish. Thin, transparent, and the color of sherry wine, dunfish was considered a great delicacy in its day.

The Isles of Shoals are rich with lore of pirates. Blackbeard, Scott, Kidd, and Quelch are reported to have landed here. According to legend, Blackbeard brought one of his many wives to honeymoon on Smuttynose. When the British fleet was seen on the eastern horizon, Blackbeard escaped, leaving his wife to guard his treasure. Since then, several people have reported seeing a ghostly figure of a fair woman, standing with eyes fixed out to sea, who cries in a mournful voice, "He will come back, he will come back, he will come back." This story is the basis of a Star Island tradition. Departing guests are given a dockside sendoff by a chorus of people saying, "You will come back, you will come back, you will come back!"

One of the most colorful chapters in the Isles of Shoals history is the development of the nineteenth-century summer resort hotel that became a popular center for artists and intellectuals of high society. Appledore House was the dream of Thomas B. Laighton, a prominent Portsmouth businessman and politician. Laighton surprised his friends by purchasing several of the islands, taking the position of lighthouse keeper on White Island, and moving there with his family in 1839. A few years later, in 1848, Laighton and Levi Thaxter opened Appledore House, a resort that grew to great fame and popularity.

Laighton's daughter Celia married Thaxter and became a well-known poet and author. Her hospitality and charm drew many admirers to the island hotel. Among the guests at the Appledore House were President Franklin Pierce and artists such as Childe Hassam and Ross Turner. Writers included Nathaniel Hawthorne, John Greenleaf Whittier, James Russell Lowell, and Sarah Orne Jewett. E. M. Statler, founder of the Statler hotel chain, reportedly visited Appledore House and picked up ideas for his hotels.

A second hotel on Star Island, the Oceanic, was opened by John Poor in 1873. This hotel burned the next year, was rebuilt by Poor, and sold to the Laighton family in 1875. Both hotels served a large clientele during their golden era of the 1880s. Business declined during the 1890s and the Laightons were forced to mortgage their Appledore holdings. In 1900 the bank foreclosed on this mortgage. The Appledore House burned to the ground in 1914.

The Oceanic survived through a stroke of good luck when it was chosen as the site of a Unitarian conference in 1897. This old hotel is still in service as the main building of the Star Island Conference Center. Started in 1897 and held every summer since then, these conferences have grown into an annual 11-week season covering religious and educational subjects.

Scientific study led to the rebirth of Appledore. In 1928, Professor C. Floyd Jackson of the University of New Hampshire established a summer marine program on Appledore, which lasted until it was forced to close during World War II.

Science programs at the Isles of Shoals resumed in 1966 under Dr. John M. Kingsbury, a professor from Cornell University. Cornell later joined forces with the University of New Hampshire to form Shoals Marine Laboratory, operated on Appledore since 1973.

How to Get There

From I-95, turn off at exit 7 onto Market Street Extension and turn toward the middle of Portsmouth. The Viking Cruises dock is one mile ahead on the left. Address: Viking Cruises, Box 311, Portsmouth, NH 03801, (603) 431-5500.

Portsmouth's waterfront is a great place to browse while waiting for the boat to leave. Walk up Ceres and Bow streets, full of small craft and gift shops, toward Prescott Park's beautiful gardens and Strawbery Banke, an area of restored homes and craft shops representing 350 years of local waterfront history.

Although Viking Cruises runs several trips a day to the Isles of Shoals, only the 11:00 a.m. trip allows stopover passengers (100 tickets only, first come, first served), with ticket sales opening at 9:00 a.m. Stopover tickets often sell out quickly, so be early. There is no charge to land on the islands. Passengers going to Star Island will have three hours, between noon and 3:00 p.m., to roam the island.

Passengers going to Appledore are requested to make prior arrangements, if possible, through the University of New Hampshire Marine Program (603) 862-2994, which is in

Student at Shoals Marine Laboratory on Appledore Island collects species from the intertidal zone.

daily radio contact with Shoals Marine Laboratory. (On certain days, the school is too busy to accept visitors who need transportation.) Take the *Viking Sun* to Star Island, then board the Shoals Marine Laboratory's open launch, which meets every Viking Cruise boat, weather permitting. It's a half-mile trip (sometimes a splashy one) across the harbor to the dock at Appledore. Yachtsmen are welcome to row ashore and land at SML's dock during daylight hours of the operating season (May to September).

Lodging

No inns or hotels operate on the Isles of Shoals. Star Island Conference weeks are often oversubscribed, with long waiting lists. Contact address: The Star Island Corporation, 110 Arlington Street, Boston, MA 02116 (617) 426-7988.

The Shoals Marine Laboratory offers, in addition to college credit courses, several three-day and five-day workshops open to the public ("for families and adults"). Typical courses might be: Science of the Sea, Nature Photography, Marine Mammals (with whale-watching boat trips), and From Sea Floor to Table (includes seafood cooking classes).

Contact address: Shoals Marine Laboratory, G-14 Stimson Hall, Cornell University, Ithaca, NY 14853, (607) 256-3717. Summer address: Post Office Box 88, Portsmouth, NH 03801, or University of New Hampshire Marine Program, Durham, NH 03824, (603) 862-2994.

Facilities for Yachtsmen

Gosport Harbor, open to the northwest and lined with a rock and kelp bottom, is not the best of anchorages. *A*

Cruising Guide to the New England Coast suggests trying to anchor "in the soft bottom, well up in the cove between Cedar Island and Star Island."

There are no facilities for yachtsmen, but sailors may land on Star or Appledore during daylight hours, from sunrise to sunset. Follow the rules for visitors and do not bring dogs ashore, light fires, or leave trash.

Meals

For Star Island, pack a picnic lunch and take it with you. Children should be supervised, particularly on the exposed outer shores, because of unexpectedly high waves that can wash a person into the ocean, with its dangerous undertow. Lunch boxes are available on the *Viking Sun* (check before you leave the dock). Remember to take your trash home with you and leave the island as clean as you found it.

Star Island Snack Bar is open daily, with sandwiches, fruit, soda, and ice cream. Since most of its patrons are conferees, the snack bar closes when conference meals are being served. Hours: 9:30 a.m. to 11:30 a.m. and 1:30 p.m. to 11:00 p.m. The Star Island dining room is open to conferees only.

Shoals Marine Laboratory has no public snack bar. The dining room is open to students and staff only, except by prior arrangement. When the school is not crowded, it is sometimes possible for a visitor to make arrangements to buy a meal.

What to Do   Star Island

A daily guided walking tour leaves from the east end of the main building porch at 1:15 p.m., going along the trails to the back cliffs and taking in the points of interest such as the monuments and meetinghouse.

The buildings open to the public are the old hotel (main floor and basement only), the meetinghouse, and Vaughn Memorial. Day tourists are requested to stay out of all other buildings.

Visitors may roam the lobby, snack bar, bookstore, and gift shop of the conference center (old hotel) and may use the rest rooms in the basement. The beautiful stone meetinghouse (chapel) is on the rise behind the hotel. The Celia Thaxter room in the Vaughn Memorial Building contains memorabilia of the area's most famous poet and author.

Visitors are requested not to go barefoot on Star and Appledore islands because of rusty nails, broken glass, and broken shells. A hat and dark glasses provide protection from the sun.

Appledore Island

All visitors should land at the dock, where they may sign in and pick up a map of the island and list of regulations.

Trails are marked. There are two basic loops: the southern trail takes a little over half an hour, and the northern trail takes somewhat over an hour, depending upon stops to look at birds or plants.

Laighton cemetery, the foundations of the Appledore House, and Celia's garden are close to the boat landing. Volunteers have re-created Celia Thaxter's garden as it was in 1894 when she wrote *An Island Garden*.

With previous permission, individuals may visit classrooms and labs. Group tours can be arranged through

Shoals Marine Laboratory buildings on Appledore Island viewed from the re-creation of Celia Thaxter's nineteenth-century garden.

the Marine Program at the University of New Hampshire, (603) 862-2994, led by one of the docents, who will meet you in Portsmouth and go with you to Appledore.

Visitors may use rest rooms at Kiggins Common, picnic in the designated area, or swim in the tidal pool (do not go barefoot).

Casco Bay ferry boats carry passengers and mail between Portland and six residential islands.

Casco Bay Islands

FIFTY MILES northeastward over open ocean from the Isles of Shoals are the broad waters of Casco Bay. Stretching 20 miles from Cape Elizabeth to Cape Small and 12 miles in width, Casco Bay is streaked with long, thin islands and peninsulas, scraped long ago into this shape by the giant claws of a glacier. These islands run parallel to each other like a school of oversized fish swimming from northeast to southwest.

They are known as the Calendar Islands because of an eighteenth-century official who reported Casco Bay had "as many islands as there are days in the year." A more careful count reveals not 365, but closer to 136 islands, still more than any other Maine-coast bay.

For those of us who love boats yet have a tendency to seasickness, Casco Bay is a dream come true, for all these islands break down the ocean swells leaving the waters more like those of a large lake.

The Islands

Most of the residential islands lie along the Casco Bay ferry route. Peaks, the Diamonds, Long, and Cliff are all part of the City of Portland, with city taxes and a minimum of city services. Children go to elementary school on their respective islands, then commute on the boat to Portland for junior and senior high school. Great Chebeague is part of Cumberland, and Chebeague teen-agers cross the water and take a bus to Cumberland high school.

Each island has its unique character. Peaks, Long, Great Chebeague, and Cliff have year-round communities; the first three offer facilities for day tourists. Starting at Portland and traveling along the ferry route, Peaks has the largest population, the shortest commute to Portland, and a general feel of a close-knit neighborhood on the outskirts of a city. Next in line are the Diamonds, populated by vacation homeowners in summer, yet almost ghost towns in winter. Long Island is truly long and narrow. It is known for its beautiful state beach, which is a mixed blessing for Long Island residents who have to contend with crowds on summer weekends and holidays. Great Chebeague (also called Chebeague Island or simple Chebeague) is the largest island, with the highest elevation, two separate ferry operations, and the only good-sized inn on a Casco Bay island. Cliff Island is the furthest out to sea, a lobstering community with a number of summer homes.

In addition to the residential islands along the ferry route are three other islands served by commercial tour boats or small island launch. Eagle Island, once the home of North Pole explorer Admiral Peary, is maintained as a museum. House Island offers an interesting fort with underground tunnels to explore as well as charter trips including group lobster bakes. Both Eagle Island and House Island

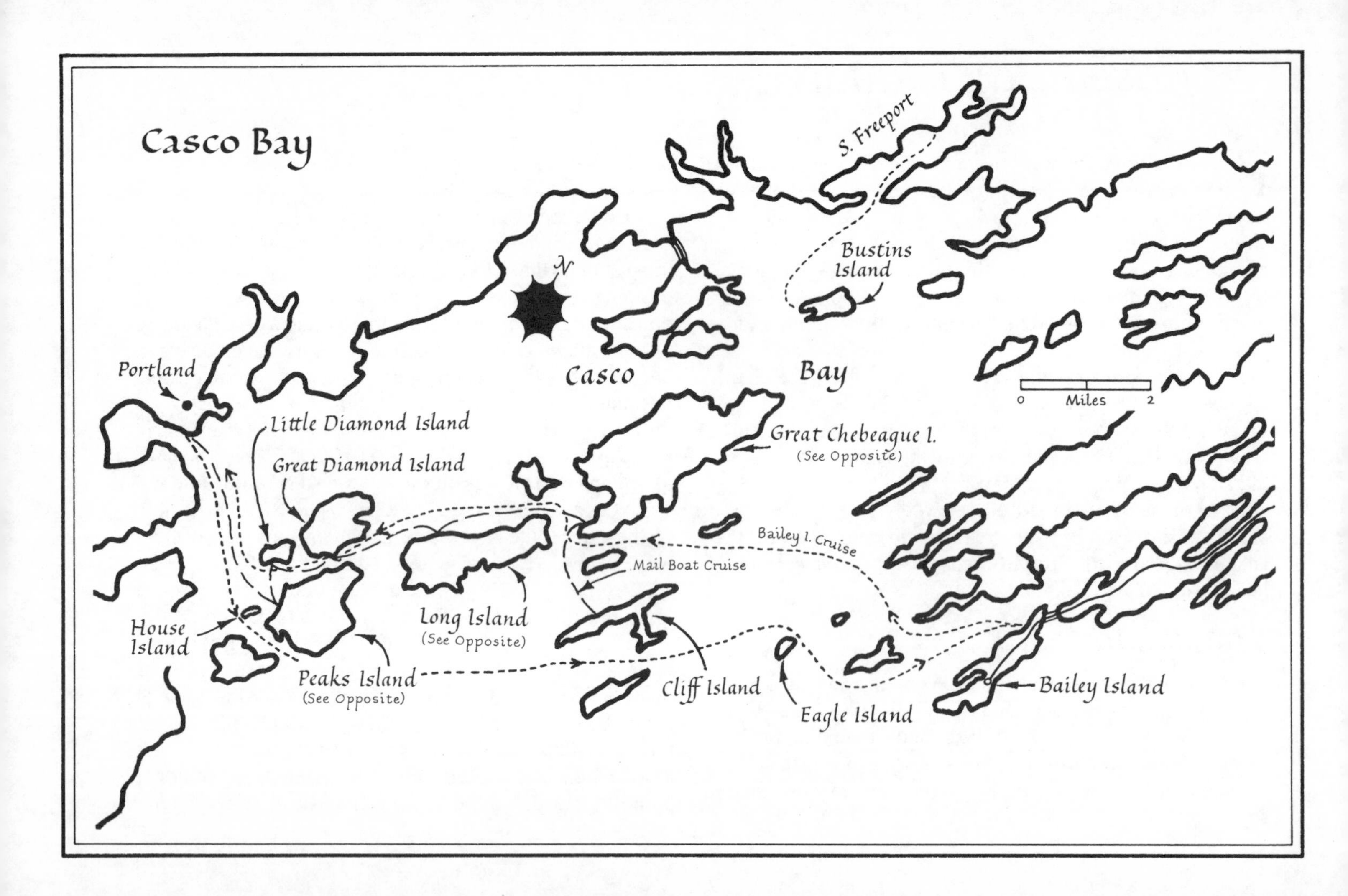
Casco Bay
S. Freeport
N
Bustins Island
Casco
Bay
0
Miles
2
Portland
Little Diamond Island
Great Diamond Island
Great Chebeague I.
(See Opposite)
Bailey I. Cruise
Mail Boat Cruise
House Island
Long Island
(See Opposite)
Peaks Island
(See Opposite)
Cliff Island
Eagle Island
Bailey Island

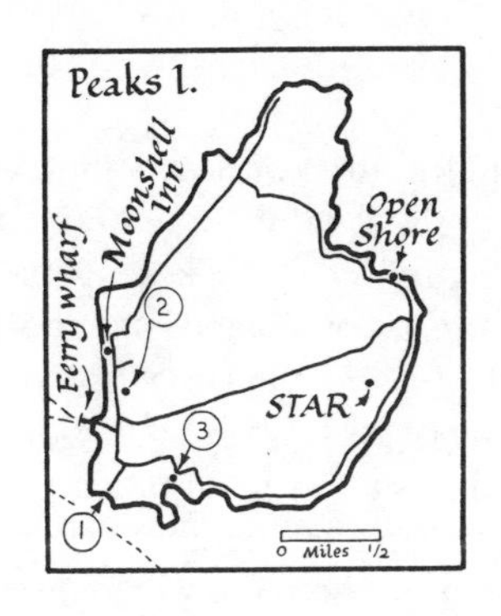

Peaks Island

1. Sandy Point Beach
2. Island Avenue shops and restaurants
3. Civil War Museum

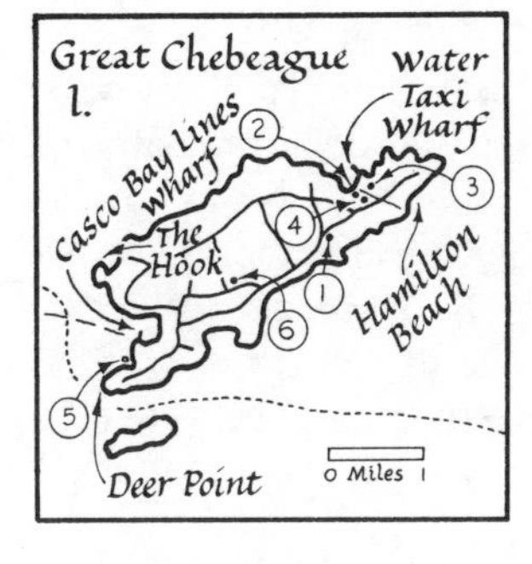

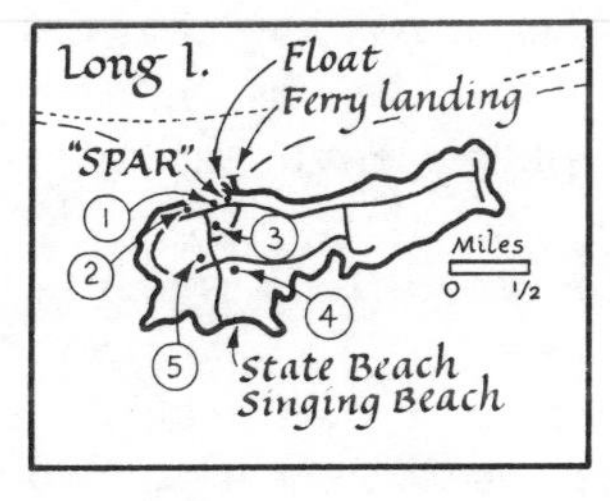

Long Island

1. General store
2. VFW Hall
3. Beach Avenue House
4. Tennis court
5. Church, gift shop, library

Great Chebeague I.

1. Island Market
2. Stone Pier Variety
3. Chebeague Inn
4. Golf course
5. The Nubble
6. Eva's Gift Shop

boats leave from Portland's Long Wharf. Bustin Island, a small residential island with private homes, is accessible by boat from South Freeport.

Casco Bay Lines

Portland's waterfront, the newly renovated Old Port Exchange, is a tourist's delight of narrow streets, craft shops, and tiny restaurants. It's easy to spot the red, yellow and black Casco Bay Lines ferryboats, the *Island Romance*, the *Island Holiday*, the *Abenaki*, or the car ferry (the *Rebel*) docked at the wharf.

Casco Bay Lines has been operating ferries in these waters since the company was chartered by the state legislature in 1845. The company is operated now by Casco Bay Island Transit District. The boats are large, carrying between 170 and 300 passengers. Besides the numerous commuter trips each day all year long, there are cruises geared especially for tourists.

The most popular cruise is the four-hour, round-trip Portland to Bailey Island ride, which weaves through dozens of islands. Refreshments are sold aboard the boat, but many people plan to get off the boat for a seafood lunch at a Bailey Island restaurant and to return on the afternoon boat. Bailey Island is accessible by car, and some people drive there to start the cruise by traveling west toward Portland. Casco Bay Lines plans to run a third Bailey Island cruise in the evening in summer.

The shortest and most frequent trip is to Peaks Island, only seventeen minutes from the Portland wharf. Peaks Island commuters and other passengers make up over half of the total Casco Bay Lines ticket sales. The ferry goes 26 times a day in summer starting at 6:00 a.m. and continuing until midnight. Even in winter, Peaks ferry runs 17 times a day. Peaks is the only island that rates a regularly scheduled car ferry.

Another popular tourist option is the twice-daily mailboat cruise that stops at Little Diamond, Great Diamond, Long, Great Chebeague, and Cliff. Passengers who would like to stop for a few hours on one of the islands—to bike around Great Chebeague or swim on Long Island — may do so, continuing their tour of the bay via a later boat. Ferries follow this route several times a day in summer.

Some people like to plan to return to Portland after dark, enjoying the evening on the ferry, watching the sunset from the deck as the first stars appear overhead. The tiny lights of houses delineate the islands you saw by day, and you may be lucky enough to see the moon rise over the water or view the *Scotia Prince*, the 410-foot ferry that sails nightly from Portland to Yarmouth, Nova Scotia.

To get to the Casco Bay Ferry Lines from the Maine Turnpike (I-95), take exit 6A and follow I-195, then signs to the Waterfront. Proceed about two miles along Commercial Street to Custom House Wharf (look for Boone's Restaurant on the same wharf). Parking is available at Custom House Wharf, or nearby Long Wharf ($3 or $4 per day). Contact: Casco Bay Lines, Custom House Wharf, Portland, ME 04101 , (207) 774-7871.

Peaks Island

There is barely time to strike up a conversation in the seventeen minutes before the boat docks at Peaks. The island is thickly settled on the Portland side and has a large area of undeveloped shorefront, open to the public, on the ocean side. Being so close to Portland, Peaks Islanders feel they have the best of both worlds. One Peaks Islander puts it this way, "We have a beautiful small-community atmosphere. We can swim in our front yard and bicycle around the island; or we can take our own boat and be in Portland in five minutes to go to the symphony."

Because of the easy commute, all kinds of people live on Peaks, and some people call it "the little city in the bay." It's popular with artists and writers, who love the relative peace of an island atmosphere, as well as with executives, lawyers, and blue-collar workers who hold jobs along Portland's waterfront. About 1,500 people live here year-round, a figure that jumps to roughly 6,000 in summer on this 720-acre island. Peaks boasts more shops and facilities than any other Casco Bay Island, with three restaurants, gift shop, supermarket, gas station, post office, bakery, laundromat, bike rental business, library, elementary school, child-care center, and two churches.

History of Peaks

The proximity to the mainland, which is advantageous today, was a liability to early settlers, who were subject to raids by Indian people who traditionally summered on these islands. A raid in 1689 discouraged white settlement for more than a hundred years.

The most colorful part of Peaks Island history started in the last half of the nineteenth century, as this island became an amusement park for Portland. It was here in 1887 that America's first summer-stock company had its start, in the Pavilion, a theatre at Greenwood Gardens.

By the turn of the century, Peaks was enjoyed by tourists who came by steamer from Boston and New York, and by Grand Trunk Railway from Montreal. Peaks Island offered several theatres, roller rinks, a shooting gallery, ferris wheel, merry-go-round, cages of wild animals, and a man who made daily balloon ascensions. For 25 cents, people could buy a round-trip ticket from Portland to Peaks including admission to the theatre.

How to Get There

Ferries to Peaks run almost hourly, from early morning until about midnight, so it is possible to have dinner on Peaks Island, then return to Portland for the night. Be wary of taking a car over on the car ferry because there may be a several-hour wait in line for the return trip. A taxi meets every boat, so a car is unnecessary.

Lodging

Moonshell Inn, run by Bunny Clark, opened in 1983, is tastefully decorated in simple, colorful style. Breakfast is fruit, cereal, and home-made muffins. Reserve ahead; inn is open year-round. Moonshell Inn, Peaks Island, ME 04108, (207) 766-2331.

Facilities for Yachtsmen

Lionel Plante Associates has a float and provides gasoline and water. Ice is available at a nearby grocery store. Moorings can be rented for a few hours or overnight. Yachtsmen who anchor in the area may land and pull dinghies up on the float for a small fee. Lionel Plante Associates, Island Avenue, Peaks Island, ME 04108, (207) 766-2508.

Jones Landing is constructing space for yachtsmen who wish to dine at the restaurant.

Meals

Cockeyed Gull has an outside deck overlooking the water, and offers breakfast, lunch, and dinner, year-round. The delicious baked goods are available on a take-out basis.

Keller's Dairy Bar specializes in pizza, lobster rolls, and Italian sandwiches, served for take-out or sit down. Wednesday through Sunday. 11:00 a.m. to 9:00 p.m. Summer only.

Jones Landing, near the ferry landing, is open year-round for lunch and dinner. It also offers a cocktail lounge and a large open deck with an outstanding view of the bay.

What to Do

Biking is good along the paved road that circles the island, which is about three miles long by two miles wide. Rent a bike from Lionel Plante Associates, near ferry landing, or bring your own.

There is swimming at Sandy Beach not far from the landing, or at one of the other beaches.

The back shore is open land, formerly a military base, saved for public access by a group of civic-minded Peaks Islanders who formed the Casco Bay Island Development Association. It's wonderful for walks and picnics.

STAR (Solar Technology and Applied Research) operates an experimental solar energy demonstration project with two solar greenhouses, organic gardens, and nature conservancy. STAR is located at Battery Steele, a former World War II 16-inch gun emplacement. STAR is open June through October. Picnics are allowed. Write STAR, Box #13, Peaks Island, ME 04108. Gretchen Hall, Casco Bay Island Development Association (see below) will give a tour by appointment.

A Civil War Museum, the Fifth Maine Regiment Community Center, is open daily during the summer.

For information on Peaks Island or on other Casco Bay islands, contact Gretchen Hall, secretary of Casco Bay Islands Development Association, Peaks Island, ME 04108, (207) 766-3381.

Great Diamond and Little Diamond

Across the bay from Peaks Island are the Diamonds, connected to each other by a sandbar at low tide. The Diamonds are part of the City of Portland, and city trucks drive across the bar at low water for rubbish collection. Both islands discourage tourists who only want to walk around; there are no public facilities, no restaurants, and no public beaches. On Little Diamond, even the roads are private. The mailboat cruise stops at each of these islands to unload and load mail and passengers.

At Great Diamond, it's not unusual to see a friendly, white-haired lady bringing a batch of warm doughnuts down to the ferry for her grandson captain and his crew.

About 250 people live here in the summer along the tree-shaded dirt roads and old wooden boardwalks, but most everyone is gone in January. Bob and Connie Noring and their two girls remain; Bob takes care of the gravel roads and keeps an eye on things.

Over half of Great Diamond's 350-acre area is taken up by Fort McKinley, a long-neglected World War II military base, now owned by Phoenix Oil Company. The No Trespassing signs at the entrance have not stopped vandals, who over the last thirty years have broken almost every window and hauled off practically every stick of furniture from the barracks. What was once a proud base for a thousand men is now grown up in high bushes and weeds, and has become a dump for the island's old junk cars. Diamond Cove, once a busy port for the boats that set submarine mines, is now a lazy anchorage for summer yachtsmen. Some Saturdays there are up to fifty yachts in this little cove.

Long before the military days, when the first settlers moved into this area, Great Diamond was used as a huge hog pen and was named Hog Island. This was common practice in those days because islands were safe for the animals, requiring no fencing to keep out the wolves. There are many Hog islands, Ram islands, Cow islands, and Sheep islands all along the Maine coast. In 1882, resort homeowners renamed this island Great Diamond, taking the name from Diamond Cove's quartz crystals, which shine like jewels in the sun.

Little Diamond's 73 acres are split, with the north end owned by the Roman Catholic Sisters of Mercy, who run a summer home for children. The southern end is a community of about 125 people in 40 seasonal cottages. The old Coast Guard Station, used during World War II as a submarine net repair operation, is the only year-round residence. It is owned by Ted Rand, who has lived here since 1954. Rand, the island constable under the City of Portland, states that the people of Little Diamond do not encourage ferry tourists to land on this island. Says Rand, "We have no public land, no sanitary facilities, no public roads; where are they going to go?"

Facilities for Yachtsmen

Ted Rand operates a fuel dock on Little Diamond for visiting yachts. He offers moorings, diesel fuel, water, ice, lobsters, and groceries.

Long Island

Long Island, three miles long and one mile wide, has 988 acres and a year-round community of about 125 people, expanding in summer to roughly 2,000 people in 298 homes. During the grand days of summer island hotels, Long Island boasted the Casco Bay House, the Dirigo House, the Granite Spring Hotel and the Beach Avenue House. Today's visitors stay in their own vacation homes or rent a house for a week or two. Long Island is popular with Portland people because of state-owned Big Sandy Beach on the south side of the island.

When I visited on a sunny weekend in August, my boat was loaded with about 100 people headed for the beach. Some stopped in for an ice cream cone at the Spar, a restaurant next to the ferry landing, where a few children played video games and customers joked with the woman behind the lunch counter.

At the grocery store down the road, proprietor Edgar Clarke showed me a scrapbook with photographs of the U.S. Navy ships anchored there during World War II, when Long Island was the refueling depot for the North Atlantic Fleet.

A quiet man in his sixties, Clarke is following in the footsteps of his father and grandfather, both storekeepers on Long Island. Clarke spoke about the change in the island since the state took over the beach: "More people are coming just for the day."

I took the taxi tour through the deserted 183-acre World War II refueling base, now owned by Phoenix Oil Company, and out toward the east end, passing picturesque Long Cove and its wharf loaded with lobster traps. We watched fishing boats off Harbor de Grace on the ocean side, and ended up at Big Sandy Beach, where people sunned on the white sand and waded into the chilly ocean. The beach is on the opposite side of the island from the ferry, about a one-mile walk, and although most of the people had come on foot, about twenty had biked over, and several others were from the three sailboats anchored in the cove.

Lodging

Beach Avenue House rentals include three large apartments plus a house. They are available by the week in summer and by the weekend during the spring and fall. Contact Mark and Linda Greene, 3 Kendall Hill Road, Sterling Junction, MA 01565, (617) 422-6293. In summer, contact the Greenes at Long Island, ME 04050, (207) 766-4440.

Meals

The Spar, open in summer only, serves lobster dinners, pizza, sandwiches, and ice cream cones. Hours: 6:30 to 7:00 a.m. (for breakfast) and 10:30 a.m. to 10:30 p.m. Sundays and Holidays: 10:00 a.m. to 1:30 p.m. and 3:30 p.m. to 10:30 p.m.

Clarke's store carries ice cream, snacks, soda, beer and other lunch fixings. The store is open year-round.

What to Do

Big Sandy Beach is designated as a "primitive" beach, with no changing house, food concessions, or rest room facilities. (Day trippers would do well to take advantage of the rest rooms on the ferry.) Be considerate of signs showing where the state land ends and private property begins. Along this shore is "Singing Beach," where the waves make an eerie noise when the wind is just right.

Biking is good, with quiet roads (some are paved) and beautiful views. There are no bike rentals, but many people bring their own. A public tennis court is available for people staying on the island.

A gift shop and library are located in the old schoolhouse, owned by the Methodist Church, open 1:00 p.m. to 4:00 p.m., seven days a week.

The VFW hall, available for weddings, etc., can be reserved through Mr. Olan Wood, Long Island, ME 04050, (207) 766-2575.

Long Island's Big Sandy Beach is a popular tourist attraction.

Great Chebeague

Great Chebeague Island (also called Chebeague Island), largest of the Casco Bay islands, is four miles long and 1½ miles wide, with a total of 2,130 acres. It's the hilliest island as well, with a high point 190 feet above sea level.

The community of 400 year-round people and about 2,000 summer residents is large enough to support a ferry on each end of the island, two grocery stores, two taxis, a boat yard, garage, golf course, gift shop, sailing school, and a good-sized inn with restaurant. Community services include post office, elementary school, church, community health center, and an emergency fire and rescue squad.

Visitors often come over to Chebeague on the convenient water taxi from Cousins Island to play a round of golf, have a meal at the Chebeague Inn, or to stay there overnight. The Chebeague Inn has fourteen bedrooms and a large living room, with fireplace, bookshelves, and games. It's a good spot for families with children. My double room was simply furnished with twin beds, braided rug, rocking chair, and a corner wash basin. Shared showers and toilet facilities were down the hall. Some rooms have private baths.

Going around the island by bicycle, I stopped at the Island Market and talked with Edmund Doughty, the 25-year-old proprietor, who recently took over the store from his father, Earle. The building that houses the store is almost 200 years old and Edmund Doughty is descended from one of the island's first families. "When you live here, you forget you live on an island," said Doughty. "It's like it becomes your whole world."

Down at the boat yard, 73-year-old Alden Brewer hauls and repairs boats—and sometimes gets embroiled in politics. Some years ago, he backed a proposal in the legislature that would have funded building a bridge from Chebeague to the mainland via Cousins Island. "We almost made it," said Brewer, who would prefer to drive to his island, rather than carry his groceries and other mainland purchases, bag by bag, onto the ferry. But the measure was turned down in a statewide referendum.

Just down the road from the boat yard, my bike blew a tire. Islander friendliness was demonstrated by the lady at the nearest house, Sylvia Hamilton Ross, who wouldn't let me call a taxi, but insisted upon personally driving me, and my injured bike, back to the Inn.

History

The name Chebeague comes from an Indian word meaning island of many springs. Abenaki Indians were the first summer people on Chebeague, fishing in the waters, digging clams, and hunting seals. Indians families have continued to summer on Chebeague up until recent times. Carroll D. Hamilton, age 90, recalls that during his lifetime Passamaquoddy Indians used to hunt seals on this island and use the pelts for clothing. An Old Town Indian family, the Atteans, used to tent near the tennis courts during the 1930s. The Atteans made and sold baskets and beaded moccasins.

White settlement of Chebeague began in earnest after Zachariah Chandler purchased land here from the First Church of Boston in 1743. Chandler had ten brothers and sisters, many of whom joined him in clearing land and building homes.

In 1760, Ambrose Hamilton bought 100 acres on Chebeague, fathered 14 children and 71 grandchildren. Many of his descendents still live on this island, including Sylvia Hamilton Ross who helped me with my bicycle. Ambrose Hamilton started what was to become Chebeague Island's most famous industry: transportation of granite in boats known as "stone sloops." By the 1850s there were about thirty of these vessels, sailing their heavy cargo from the granite quarries on Vinalhaven, Stonington and other ports to the east-coast cities of New York, Boston, and Philadelphia, where the stones were used to erect massive government buildings.

By the 1870s, summer tourism was growing in popularity on Chebeague. Around the turn of the century, there were several hotels including the Hamilton Hotel, the Summit House and the East End Hotel, all taking guests who arrived by steamboat.

How to Get There

You can take the Casco Bay Lines, Custom House Wharf, Portland, ME 04101, (207) 774-7871 to land at Chebeague's Chandler Cove on the west end (1¼ hours).

As you approach Chebeague, you will see a huge, lone oak tree exposed to the winds and weather on Indian Point. This tree was in bad shape when Ellis Ames Ballard bought land and built an estate here in the early 1900s, but Ballard hired experts to doctor the oak, which has since survived several hurricanes.

The water taxi, *Polly-Lin II*, from Cousins Island is quicker, only fifteen minutes, and takes you to the stone pier on the east end of the island, next to the golf course and Chebeague Inn. Contact Chebeague Transportation Company, Chebeague Island, ME 04017, (207) 846-3700, for schedule. Although the boat trip is quicker, the catch is the parking. Passengers must park at Cumberland School. To find the school, take exit 9 off Route 95 and follow Route 1 north three miles to Tuttle Road. Turn inland and go northwest 2¾ miles to Drowne Road and Cumberland Elementary School. A school bus transports visitors (half-hour ride) across the bridge to the Cousins Island water-taxi terminal.

Lodging

The Chebeague Inn, Chebeague Island, ME 04017, (207) 846-9634, is open from Memorial Day to Columbus Day, serving breakfast, lunch, and dinner. The cocktail lounge and dining room overlook the water.

Facilities for Yachtsmen

The stone pier on the north end of the island has several floats. There are no facilities at this wharf for fuel or water, and the water is shallow. Docking is limited to 30 minutes, but dinghies may be tied to the outermost float. A small grocery store, the Stone Pier Variety, is located at the wharf. There is a public telephone booth near the store.

Chebeague Island boat yard is located in the center of the east side of the island. Depth of water is two feet at low tide, six to eight feet at high tide. Gasoline, oil, and water are available, (207) 846-4146.

A popular anchorage is off Little Chebeague Island south of Indian Point. The island is owned by the State of Maine. The small sandy beach makes a nice picnic spot.

Many people bring bikes on the water taxi from Cousins Island to Great Chebeague.

Meals

The Chebeague Inn's dining room is open to the public. It features a full menu at breakfast, lunch, and dinner. Many people come by boat, anchoring off the Stone Pier or taking one of the Chebeague Inn's moorings. Boats with no dinghy may borrow the Chebeague Inn's dinghy to row to shore. The Bounty Pub, in the basement of the inn, serves drinks in the afternoons and until late evening, seven days a week, mid-May to mid-October. For details, phone Chebeague Inn, (207) 846-9634.

Pick up lunch fixings at the Island Market (summer hours: 9:00 a.m. to 5:30 p.m. and 7:00 p.m. to 9:00 p.m.) or at the Stone Pier Variety, if it is open.

What to Do

The nine-hole golf course is described as a "good, challenging little course." Open to public.

For swimming, try Deer Point or Hamilton Beach, near the Chebeague Inn. The Nubble is a fine picnic spot.

Little Chebeague Island, south of Indian Point, is accessible by walking over the sandbar at low tide. For about two hours on either side of low tide the bar is above water. Little Chebeague is uninhabited and owned by the State of Maine. Leave from Indian Point on Great Chebeague, taking the public right-of-way road to where the sandbar begins. Walk across, explore the island, or picnic on the sandy beach. Be careful to keep an eye on the tide to avoid being stranded.

Biking is popular. Inn guests can rent bikes; others should bring their own bicycles.

Eva's Gift Shop, run by Eva Conrad, is open daily, 10:30 a.m. to 5:00 p.m., and Sundays from 11:30 a.m. to 3:00 p.m.

Cliff Island

Cliff Island stands on the outermost edge of Casco Bay, where the bay meets the open sea. The island is formed like the letter H. The northern side is about two miles long, connected to the bluffs on the shorter southern side by low land. Cliff Island is part of the City of Portland, and has a year-round community of about 80 people, expanding to about 300 in the summer.

Cliff Island, outermost of the residential islands along the ferry route, has a small fishing community and a number of vacation homes.

After an hour and a quarter on the ferry, I disembarked at Cliff Island and read a large sign on the ferry dock that tells about the community and explains that the island has no inn, no restaurant, no campgrounds, no public beaches, and no parks. The general store had closed for the afternoon, and the very small eating place on the island was not open until evening. I took a taxi tour past the tennis court, Association Hall, and library, then over to the boat cove on the southeast side of the island, where most of the fishermen live. We drove by the nature preserve and Kennedy's Beach, and stopped near the post office and school.

The story of the Cliff Island school is typical of schools on small, remote Maine islands. Johanna von Tiling, who taught here for about ten years in the one-room schoolhouse, told me, "The year-round population is getting smaller and the summer population is getting bigger." She last taught in 1978 when the school had 18 children in grades one through eight. There were four students in 1983–84, but three of these were the teacher's children. The state, requiring a minimum of eight children, may have to close the Cliff Island school.

"This is a very serious situation," says von Tiling, "If the school closes, we'll never get it back again, and no young families will ever come here again." In the early 1970s, a similar school closing threat put Cliff Island on national television and in *Time* magazine, but that crisis was averted when a family with six children moved to the island.

At the same time, this beautiful, remote, quiet island is beloved by people who have summered here since childhood. As their children and grandchildren have grown up and brought their families, the summer community has steadily increased, buying up land and some of the year-round homes for summer use.

City taxes are high, and many vacation homeowners rent out their cottages for part of the season. Inquiries regarding summer rentals may be made through Mrs. Eleanor Cushing, who runs the post office, (207) 766-2051.

There are many legends about pirates on Cliff Island. Captain Keiff, so the story goes, hung a lantern on his horse's neck and rode up and down the island, leading passing ships to run aground on the island's rocks so he could scavenge the shipwrecked cargo. Keiff's treasure is said to be buried on nearby Jewell Island, now owned by the state.

Meals

In the building next to the ferry landing, hot dogs, crabmeat and lobster rolls, chips, soda, and ice cream cones are available from Memorial Day to mid-September, seven days a week, 10:00 a.m. to 4:30 p.m. and 6:00 p.m. to 9:30 p.m.

Eagle Island

Eagle Island, small and privately owned, was once the home of Admiral Robert E. Peary, the first man to lead a successful expedition to the North Pole. The island is now a museum, preserved as it was when Peary lived there.

The house is situated on the northeast point of the island, looking north, appropriately enough. A huge, round, stone cistern that could store up to 40,000 gallons of rainwater has porthole windows salvaged from an old ship. In the living

North Pole explorer Admiral Robert E. Peary's former summer home on Eagle Island.

room are mounted arctic birds, many stuffed by Peary himself, and a fireplace built with Eagle Island stones and arctic quartz crystals.

Peary, born in 1856, purchased Eagle Island in 1877, finally building a summer home for his family there in 1904. Eagle Island became his principal residence, where he kept the papers and plans for his seven expeditions. After his success in reaching the North Pole, and the official recognition in 1911, Congress raised Peary's Naval rank to admiral.

How to Get There

The *Kristi K.*, a tour boat out of Portland, makes two trips daily, at 10:00 a.m. and 2:00 p.m., mid-June through Labor Day. It's a scenic trip (with narrated tour) through the islands of Casco Bay, taking 3½ hours, with one hour on Eagle Island. Captain Wayne Selberg, RR 1, Box 184, Yarmouth, ME 04096, (207) 774-6498 or (207) 846-9592.

Dolphin Marina, Basin Point, South Harpswell, ME 04079, (207) 833-6000, rents small boats and will arrange to have someone take you out to Eagle Island for an hour or two. It's a ten-minute boat ride from South Harpswell.

What to Do

Tour the museum (free admission), take a walk through the woods trails and enjoy a picnic on the lawn or beach near the Peary home. Floats, moorings, and dinghies are available for private boats.

House Island

House Island, also small and privately owned, is open for lobster cookouts and tours of the island's Civil War–era Fort Scammel. The underground fort has long dark passageways, a circular stairway, and huge stone rooms. The neatly spaced cannon ports offer views out to the bay. Although this fort was never used in battle, it was fully fortified in 1862 with 71 cannons and during World War II the fort held two 90-millimeter antiaircraft guns.

House Island is one of several islands possibly occupied in 1623 by Christopher Leavitt, the first settler in Casco Bay. Leavitt left ten men on an island and sailed to England, planning to return, but never did.

At present, there are only three dwellings on House Island, all built in 1907 by the Public Health Service and used as a U.S. Quarantine and Immigration Station until the 1930s. The present owner, Hilda Cushing Dudley, has been running tours here for thirty years and operates a charter service taking groups to the island for lobster bakes and tours of Fort Scammel. She is planning to expand her operation to accommodate individual tourists, offering a daily trip to the island, leaving Portland daily at 11:00 a.m. and returning at 2:00 p.m.

How to Get There

Captain Harold Cushing skippers the 60-foot *Buccaneer* that docks at Long Wharf, off Commercial Street at Portland's Old Port Exchange. Buccaneer Lines, Box 592, Portland, ME 04112 (207) 799-8188.

Bustin Island

Small Bustin Island lies three miles off South Freeport. It has about 100 summer cottages and roughly 350 people. In contrast to other residential islands, Bustin has no electricity and no telephones, which is exactly how the people want it.

No vehicles are allowed on Bustin Island except for a few trucks to carry luggage and maintain homes. No taxi meets the boat; on Bustin Island, you walk. But it's not a long walk. One can stroll around the whole island — about two miles—in 40 minutes. The island has no sandy beaches and no place to buy lunch, but you can bring a picnic and eat it at the picnic tables near the boat landing. Islanders have no objection to someone quietly visiting their island, as long as the visitor is respectful of private property, and careful not to leave trash behind.

How to Get There

From I-95, take exit 17 to South Freeport. Follow the road down to Harraseeket Harbor. A 36-foot boat carries 34 passengers on the 20-minute ride to Bustin Island, running seven days a week, three to five trips a day, from May until the end of October.

From Fort Scammel passages on House Island, you can look out to the waters of Casco Bay.

View from Monhegan across the harbor to the island of Manana where the United States Coast Guard operates a fog signal station.

Monhegan

A SMALL CHUNK of land in a large blue sea, Monhegan stands ten miles off the nearest mainland. Wind surf pounds upon its dramatic cliffs, which drop a sheer 150 feet into the sea. When fog wraps the island in a giant white blanket, Monhegan becomes a land unto itself.

The island is blessed with beautiful woods and wildflowers—as well as with delicious berries, appreciated by birds and people. Monhegan offers a much-needed resting place for thousands of migrating birds in spring and fall.

Monhegan (pronounced "Moneeg'n" by local residents) is a good size for visitors, small enough to walk around easily yet large enough to support a thriving year-round community. The economic base is tourism in summer, lobstering in winter. In 1983 the year-round population was 85, and the number increases to about 850—plus day tourists—in the summer. A one-room schoolhouse for kindergarten through eighth grade had eight children in the 1983-84 school year.

Monhegan has a well-known artists colony, active over several generations. Artists, intellectuals, and writers give the island a bohemian ambiance. Many of the lobstermen are sons of summer people and come from a cosmopolitan background. More than any other Maine island not bridged to the mainland, Monhegan is a tourist haven. Up to 100 trippers arrive daily to browse through the gift shops and walk the trails.

First Impressions

The Monhegan mailboat, *Laura B.*, out of Port Clyde, takes up to 90 passengers twice daily in summer for the one hour trip to the island. The sixty-five foot, eighty-ton vessel was built as a T-Boat during World War II and has the sturdy appearance of a rugged fishing vessel. A sternhouse protects passengers in cold or wet weather, and a mast with cargo boom is used for loading heavy baggage.

On a typical summer crossing, the ample foredeck is piled with suitcases, boxes, duffle bags, cases of beer, cartons, windows, a water heater, bottled-gas tanks, and a small load of firewood. Sometimes the deck is loaded with chickens, a pig, or even a large pick-up truck. In nice weather the passengers congregate on the foredeck, enjoying the sun and breeze, reading, chatting, and watching the gulls, eiders, or cormorants. The *Laura B.* often startles seals, sunning on some lonely rock, who slide down the seaweed and plunge into the icy waves.

A surge of excitement ripples through the passengers as the mailboat approaches Monhegan's picturesque harbor filled with fishing boats moored closely together. Fish houses line the shore. The land rises gently, with wooden buildings on every knoll like seats in an amphitheater. Dramatically situated atop the highest peak stands the lighthouse.

Across the harbor lies the tiny island of Manana, whose

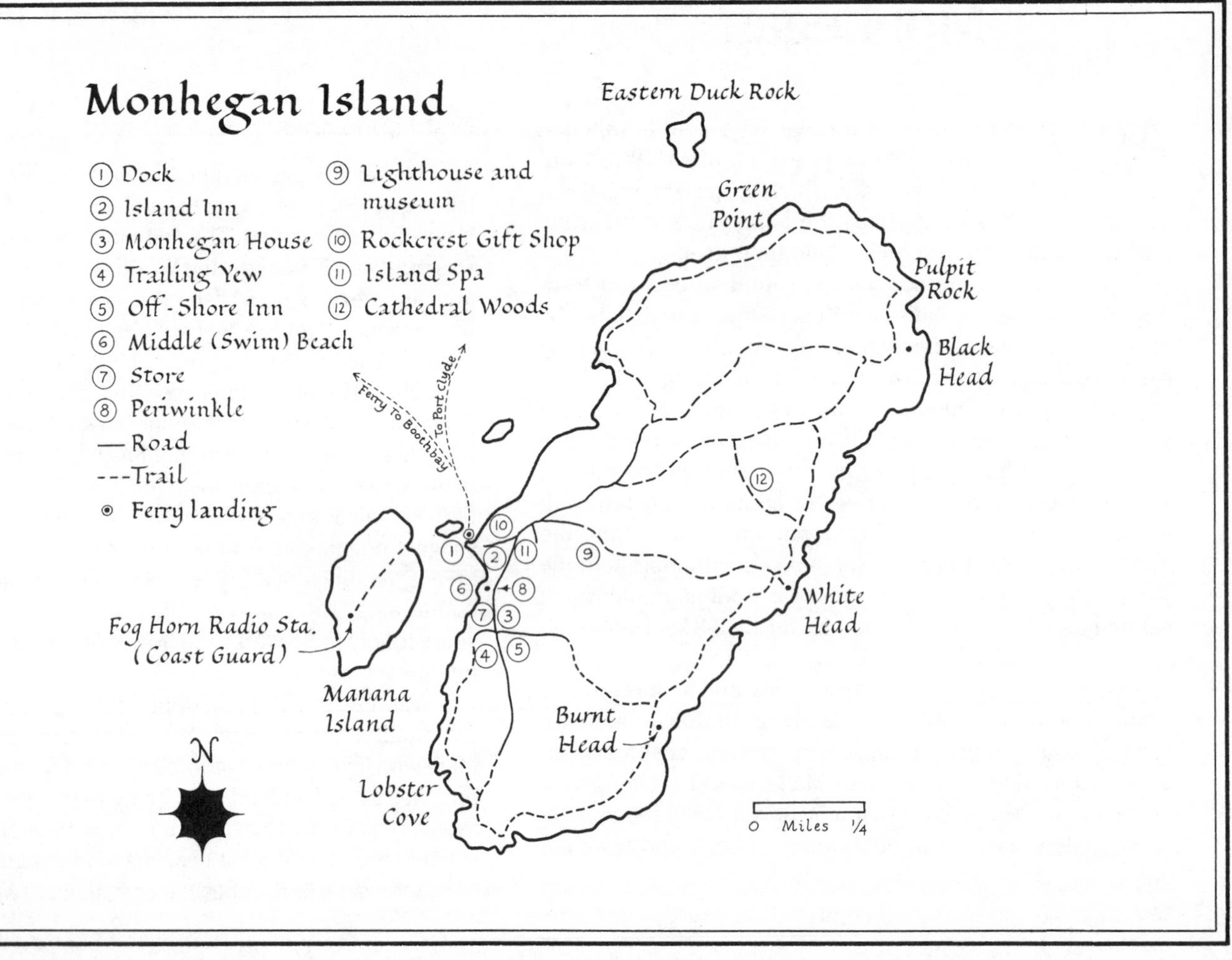

Monhegan Island
Eastern Duck Rock
① Dock
② Island Inn
③ Monhegan House
④ Trailing Yew
⑤ Off-Shore Inn
⑥ Middle (Swim) Beach
⑦ Store
⑧ Periwinkle
⑨ Lighthouse and museum
⑩ Rockcrest Gift Shop
⑪ Island Spa
⑫ Cathedral Woods
— Road
--- Trail
Ferry landing
Green Point
Pulpit Rock
Black Head
White Head
Burnt Head
Lobster Cove
Manana Island
Fog Horn Radio Sta. (Coast Guard)
Ferry To Boothbay
To Port Clyde
N
0 Miles 1/4

grassy fields were kept open by sheep until about 1980. It was here that much-photographed hermit Ray Phillips used to sleep in the same room with his sheep. Phillips died in the mid-1970s. The present occupants of Manana are the Coast Guardsmen who tend the fog signal station, plus one or two other residents.

Luggage trucks, but no taxis, meet the boat at the Monhegan wharf. On this island, vehicles are prohibited except for a few necessary trucks. It's wonderful to walk along roads without cars or exhaust fumes. The half-mile stretch of narrow dirt road that is Monhegan's main street winds from schoolhouse hill past the library, news store, post office, grocery store, church, and several inns.

Monhegan's electricity comes from individual private generators or solar modules, and two or three of the inns use kerosene lamps in the bedrooms. Town water is available in summer only, gravity-fed from two high towers through pipes laid over the ground. For over twenty years, telephone service has been limited to a few pay phones using a World War I underwater cable that gradually disintegrated. A new radio microwave system made private telephones available in 1983.

Monhegan residents do not want the part of the island outside of the village modernized too much. They are blessed with a large tract of wild land that is protected from development. Theodore Edison, son of Thomas Edison, had the foresight to start this conservation effort in 1956. Edison spearheaded the group that formed Monhegan Associates, a trust to which Edison gave a large section of the island. Other land has been added by donation or purchase so that now the whole magnificent center woods and dramatic highlands along the back shore are under the care of Monhegan Associates.

In Cathedral Woods, tall, lean trunks of spruce and balsam reach up a hundred feet or more to close the forest over with a roof of branches. The forest floor is a fairyland of brilliant green moss, tiny mushrooms, and baby spruce trees.

Trails lead to Black Head and White Head, where cliffs drop 150 feet into the swirling ocean below. Beware of climbing down these cliffs as the rock is crumbly and the wet rocks near the water's edge can be very slippery. Unexpectedly high waves come without warning and can wash a person into churning water with dangerous undertows.

Monhegan's dramatic scenery and quiet isolation have attracted artists such as Rockwell Kent, George Bellows, Paul Henri, Andrew Winter, and James Fitzgerald. Today, so many artists work here that a schedule of the hours for each studio is made available for visitors. Among contemporary artists are Jamie Wyeth, John Hultberg, and Charles E. Martin (who draws *New Yorker* covers), all with national reputations.

At her studio overlooking Deadman Cove, Ruth Boynton showed me how she works on four paintings simultaneously, depending upon the light; with morning sun, with morning shade, with afternoon sun and with afternoon shade. "I never get tired of this place," says Boynton, whose paintings reflect the rocks, woods, and coves around her home. Boynton's family is closely tied to Monhegan, having summered here since 1948. Her son David was featured in a children's book, *David and the Seagulls.* Two sons live on Monhegan year-round; Willard is a carpenter and town assessor (selectman) and Douglas is a lobsterman.

Monhegan's lighthouse and microwave tower.

Lobstering is a wintertime pursuit on Monhegan, the only Maine island to have a closed legal lobster season. Within a two-mile radius of the island, lobsters may be trapped between January 1 and June 25 only. Trap Day, a major event of the year, is January 1 (or as soon after as weather will permit). On that day all Monhegan lobster fishermen set out their traps. No one gets a head start on the others. Lobstering in the winter may be cold work but that's when the market price is the highest. Lobsterman David Boegel reports, "We have to pound off the freezing spray with a hammer." But he denies being cold out there in the winter winds, and says, "I'm hot all day long."

In summer, Boegel and his wife run a pizza take-out restaurant, turning to the tourist trade as do many of the lobstermen. From June 25 until New Year's Day, Monhegan's main street is lined with tall stacks of lobster traps awaiting the next season. They make great bean poles for island gardeners.

Spring and fall are migration seasons not only for the birds but also for the flocks of bird watchers that gather on the *Laura B.* to journey across to the island. When I visited in September, groups from both Maine and Rhode Island Audubon Societies were on Monhegan, and every other person I met had binoculars. Leaders said it is possible to spot between 50 and 100 species of birds in a single weekend. Bird watchers rise at dawn to wander near the marsh or walk to the cliffs. I followed a group to Black Head, where we watched a sharp-shinned hawk. "It's a sharpie, all right," said the leader, peering through his telescope. "Short wings, long tail." Down in the woods we spotted a merlin. I never did see the peregrine falcon that was the talk of the dinner table at the Trailing Yew that night.

History

Monhegan has long been a landmark for sailors. Many explorers approaching the unknown New World noted this whale-shaped island with its eastern cliffs and headlands rising out of the sea. One legend says that an Irish monk, St. Brendan, who explored the Atlantic Ocean in 565 A.D. seeking the Isles of Saints, landed on an island with steep, scarped sides that could have been Monhegan. Norsemen may have sailed here around 1000 A.D., leaving markings on the rocks at Manana that some experts think are Viking inscriptions. The source of these marks is still in question; some say they might be Phoenician in origin, others explain them as mere glacial scrapings.

Monhegan was recorded by many of the early explorers, including Verrazano in 1524, David Ingram in 1569, and Martin Pring in 1603.

During the spring of 1614, Captain John Smith landed at Monhegan with two ships and set up a base for fishing and exploration. Most of Smith's men stayed at Monhegan to catch, dry, and salt codfish to bring back to Europe. Captain Smith and a crew of nine men sailed a small boat along the New England coast from Mount Desert Island to Cape Cod, creating a chart so good that it could almost be used today. Loaded with fish, furs, and information, Captain Smith's two boats returned to Europe and put Monhegan on the map.

Permanent settlement on Monhegan was established by 1618, and a community gradually developed until the French and Indian Wars. When Indian warriors under French direction set fire to Monhegan in 1689, all the island residents moved elsewhere.

Shortly after the American Revolution, Monhegan was resettled by three young men, Henry Trefethren, Francis Horn, and Josiah Starling, who brought their families to the island. Some of their descendants own parts of the island today.

Since that time Monhegan has been a fishing community, with lobstering as the economic base in recent years. Monhegan lobstermen themselves requested the state legislature to set the closed season on lobsters, formally made law in 1909.

The artists colony dates back to the early 1900s when Robert Henri moved here and influenced his pupil Rockwell Kent to join him.

How to Get There

The *Laura B.*, Monhegan's mailboat, leaves Port Clyde twice daily in summer, once daily in spring and fall, and three times a week in winter. The trip takes about an hour. Reservations are strongly suggested, particularly during the busy summer season. It is possible to take the 10:00 a.m. boat out and stay on Monhegan from 11:00 a.m. until the afternoon boat leaves at 4:00 p.m. Parking at Port Clyde is available near the dock. For information, contact Captain James Barstow, P.O. Box 238, Port Clyde, ME 04855, (207) 372-8848.

The drive from Portland to Port Clyde along U.S. Route 1 is scenic, passing through such pretty villages as Wiscasset and Thomaston. Take Route 1 to Thomaston, turn south on route 131 to Port Clyde. By air, take a commercial airline to

Fishermen haul a dinghy up above high water mark on Monhegan's Fish Beach.

Owls Head (near Rockland) and a taxi to Port Clyde.

From Boothbay, the excursion boat *Balmy Days* leaves daily at 9:15 a.m., arriving at Monhegan at 10:50 a.m. (June 15 to September 15). Passengers have four hours to explore the island before the return trip at 2:55 p.m., which arrives back in Boothbay at 4:30 p.m. For information, call Bob Campbell, P.O. Box 102, Boothbay Harbor, ME 04538, (207) 633-2284. Reservations are suggested.

Directions to Boothbay Harbor: Take scenic coastal Route 1, turn south on route 27 between Wiscasset and Damariscotta, and follow signs to Boothbay Harbor. Park at Chimney Pier or Sample's.

Sherman Stanley operates a water-taxi service, by charter only, between Monhegan and Boothbay or Port Clyde. Contact him at Monhegan, ME 04582, (207) 596-0326.

Lodging

The Island Inn is the grandest hotel on Monhegan, set on a hill overlooking the harbor, providing (unusual on this island) electric lights in the rooms, some private baths, and a good restaurant. Reservations are taken from January 1 (some dates fill up by early spring). The inn caters to couples or singles and is not so good for families with children. Contact Bob Burton, Island Inn, Monhegan, ME 04852 (207) 596-0371.

The Monhegan House is a large Victorian inn on the main road near the harbor. It has about 45 rooms, each illuminated by kerosene lamps. Proprietor Victor Lord calls his inn "a step back in time." Bathroom facilities are arranged on a special corridor, with a long row of toilet rooms and shower rooms. No meals are provided. Open Memorial Day to Columbus Day. Contact: Victor Lord, Monhegan House, Monhegan, ME 04852 (207) 594-7983.

The Trailing Yew has room for about 60 people spread out over several annexes that have the simplest of accommodations—kerosene lamps and shared bathrooms. Meals are wonderful, served family-style, and cooked by proprietor Josephine Day, who has owned and managed this inn since 1926. Among her past guests was Theodore Edison, responsible for giving so much land to Monhegan Associates. While cutting up onions and tomatoes in the kitchen, Josephine smiles and says, "I've been here all my life and island people get to be like family."

The Trailing Yew feeds not only its own clients, but people staying at other inns such as the Monhegan House and the Off-Shore Inn, and often has two sittings at meals, serving up to a hundred people. Contact Josephine Day, Trailing Yew, Monhegan, ME 04852 (207) 596-0440.

The Off-Shore Inn is tiny, with only three rooms, kerosene lamps, one shared bath, and no meals. Contact Sarah Burnham and Lexi Krause, Off-Shore Inn, Monhegan, ME 04852, (207) 594-2321.

The Hitchcock House is a small guest house, accommodating eight people, open year-round. Two of the rooms have cooking facilities. A self-contained cabin is also available. Write to Barbara Hitchcock, Hitchcock House, Monhegan, ME 04852, (207) 594-8137.

Tribler Cottage has both housekeeping apartments and rooms. Contact Richard Farrell, Tribler Cottage, Monhegan, ME 04852 (207) 594-2445.

Facilities for Yachtsmen

A Monhegan fisherman does not like to come home from a hard day's work and find a yacht on his mooring, but it happens all too often. As harbormaster Steve Rollins puts it, "It's like me driving my truck into your garage." Furthermore, the wind and tide conditions make it necessary to tie up with special procedures in order to avoid chafing damage.

The suggested procedure is to land briefly at the mailboat wharf—at the sides of the wharf and not along the end where the mailboat ties up. Drop off a member of your crew to locate the harbormaster by inquiring at the Island Spa or the Island Store. The harbormaster is usually nearby, particularly between 5:00 p.m. and 6:00 p.m., when he tries to be on the wharf. Rollins stresses that yachtsmen should not listen to advice from people standing on the dock, but should talk directly with the harbormaster or else a fisherman on his own boat. Rollins rents two moorings of his own and knows if there are others available.

Boats entering the harbor from the south should hug the Manana side of the harbor to avoid a reef that makes out from the Monhegan shore. There is plenty of water at the mailboat wharf.

The most popular anchorage is north of the wharf between Nigh Duck Rock and the mainland. Dinghies should not be tied to the wharf ladder because wind and tide can cause the bow to catch in the ladder, swamping the dinghy. Beach your dinghy at Swim Beach, the first beach south of the wharf, and tie up to a chain or rock. Many people make prior arrangements with the harbormaster before coming to Monhegan. Contact Steve Rollins, Monhegan, ME 04582, (207) 594-9585. Rollins's boat is *My Three Sons*, and his C.B. radio call letters are WXY 8789 (or ask the marine operator for "Harbormaster Monhegan").

Meals

The Island Inn serves breakfast, lunch (à la carte) and dinner at individual tables in an attractive dining room decorated by island artists' paintings.

The Trailing Yew serves breakfast, bag lunch, and dinner.

North End Pizza take-out is run by Katy and David Boegel at lunch and suppertime until 9:00 p.m. from Memorial Day to Labor Day. (207) 594-2191.

The Periwinkle Coffee Shop near Fish Beach offers meals, baked goods, snacks, and lobsters that can be eaten at inside tables or outdoor picnic tables. Open from mid-June until Labor Day. Proprietor Dint Day is the husband of Josephine Day at Trailing Yew. (207) 594-5546.

Steve's Fish House Franks offers frankfurters from a roadside wagon.

Monhegan Store, Inc., has fresh vegetables and fruits, Breyer's ice cream, wonderful cheeses, meats, and imported wines, plus a popular deli serving Italian sandwiches and hot take-out dinners such as spinach pie, ratatouille, or lasagna.

Picnickers are requested to take their trash off with them when they leave. With so many visitors each day, it is particularly important that each person help to keep the island clean.

What to Do

Walks along the trails are beautiful. People mosey

through Cathedral Woods or hike over to Black Head or White Head on the eastern shore. From these high cliffs it is possible to see Isle au Haut, Matinicus, Criehaven and Matinicus Rock. Children should be carefully supervised on high cliffs and also on lower rocks near the water's edge because of the dangerous undertow along this side of the island as far south as Lobster Cove.

Swimming is safe but cold at Middle (Swim) Beach in the harbor area.

Lobster Cove has two wrecks; the larger one is the tug *D. T. Sheridan*, which struck the rock known as the Washerwoman in a February snowstorm in 1948. The tug had been abandoned northeast of Monhegan when a fire broke out on board.

Deer are sometimes seen at dusk along the road between the harbor and Lobster Cove.

Monhegan's museum is situated in the former lighthouse keeper's house high on the hill next to the lighthouse. It's one of the best island museums in Maine, with a photographic exhibit of 209 Monhegan wildflowers and 155 birds sighted on the island, rooms full of Monhegan history, and an exhibit showing how lobsters are caught. Every summer, paintings by a different Monhegan artist are featured on the museum walls, using work of deceased artists only. Open daily from July 1 through Labor Day, 11:30 a.m. to 3:30 p.m.; Labor Day through September 15, open 11:30 a.m. to 1:30 p.m. Admission is free but donations are appreciated to help cover costs.

Manana Island, across the harbor, can be reached by making arrangements with an island boy at Middle (Swim) Beach. You can see the rock markings that some claim are Norse or Phoenician inscriptions. The Coast Guard keeps a fog signal station here as well as the generator for Monhegan's lighthouse.

The library has many books about the island, and each children's book has a bookplate inscribed: "Jackie and Edward's library to the children on Monhegan," referring to Jacqueline Stewart Barstow, 10 years old, and Edward Winslow Vaughn, 14, two children who were washed off the rocks and drowned August 10, 1926.

Art work may be viewed at Plantation Gallery across from the Periwinkle Coffee Shop.

Monhegan has several gift shops, including Rockcrest near the ferry landing, the Island Spa at the end of Dock Hill Road, and a crafts shop next to Trailing Yew Annex.

Winter Works is a Monhegan islanders' cooperative shop marketing crafts made on the island during the winter: pottery, weaving, knitwear, photography and paintings. The shop is located across from the Island Inn, open 11:00 a.m. to 1:00 p.m.

Sherman M. Stanley takes fishing charters and birdwatching groups in his pilot boat *Phalarope*. Stanley operates out of Fish Beach. Contact him at (207) 596-0326.

Schoolhouse hill is a great place to view the sunset. In the school yard the Tercentenary Tablet commemorates Captain John Smith's voyage to Monhegan in 1614.

Larry Cooper leads wildflower walks on the second Wednesday of June, July, and August, starting from the schoolhouse at 1:30 p.m. (Rain dates: the Friday following.) There is no charge, but donations to the Ecology Committee are accepted.

Helga Homere also leads nature walks during July and August. Inquire at the Island Spa for dates and times.

Matinicus schoolboys such as Jeremy Van Dyne, 11, lobster out of small skiffs.

Matinicus

MATINICUS IS THE most remote of Maine's resident island communities. The 23-mile trip from Rockland Harbor to the landing at Matinicus is long and can be rough in high seas.

The Maine State Ferry Service runs a trip to Matinicus once a month, primarily for vehicles, heavy equipment, and island supplies. When Matinicus fishermen need to go to the mainland, they often go in their own boats. The mail plane from Owls Head, near Rockland, is the easiest and least expensive way for a tourist to reach Matinicus in 1984. In spells of foggy weather, a Matinicus man, Albert Bunker, carries the mail and passengers on his lobster boat.

On this unprotected island, winter gales can be fierce. Matinicus is still recovering from the February storm in 1978 that washed away a large wharf—complete with restaurant—and undermined other buildings. A good time to visit Matinicus is in August or early September, late enough in the year to have a minimum of fog but before cold weather sets in.

The island is oval with the long axis running north and south. The length measures 1¾ miles, the width is ¾ mile. Matinicus is relatively flat and cannot be seen from more than about ten miles away. Its high point rises only 103 feet above sea level.

The island's 800 acres include thick stands of spruce and fir, rocky headlands, bogs and marshes, two small ponds, several fine sand beaches, and extensive open fields. All these make wonderful habitats for the many birds that live here and other birds that migrate through. Naturalists have counted 750 varieties of wild flowers and plants. In the fall, it's possible to sample an amazing number of different varieties of apples from Matinicus trees.

Most of the year-round residents live on the gravel road that runs down the center of the island, protected from the bitter winter winds, or on the side road that leads to the harbor. The year-round population numbers around 45, sometimes dropping as low as 32 in midwinter. Many of the islanders travel to the mainland for the worst of the winter, some going to Florida, some traveling only as far south as Portland. During July and August there may be as many as 150 people on the island: natives, summer people (homeowners), summer renters, seasonal fishermen, and a very small number of tourists.

Matinicus islanders are an independent lot. They have to be. When the generator breaks down in a blizzard no one is going to come out from the mainland to fix it. Matinicus residents do not miss the crowds that inundate other islands in the bay. "We don't want to be another Monhegan," they say. They have very few facilities for tourists — no restaurant, no inn, no gift shop, no bikes to rent, and no taxi. This is a no-frills working community, based on lobstering and run by the local islanders.

For my first visit to Matinicus, I crawled into the co-pilot

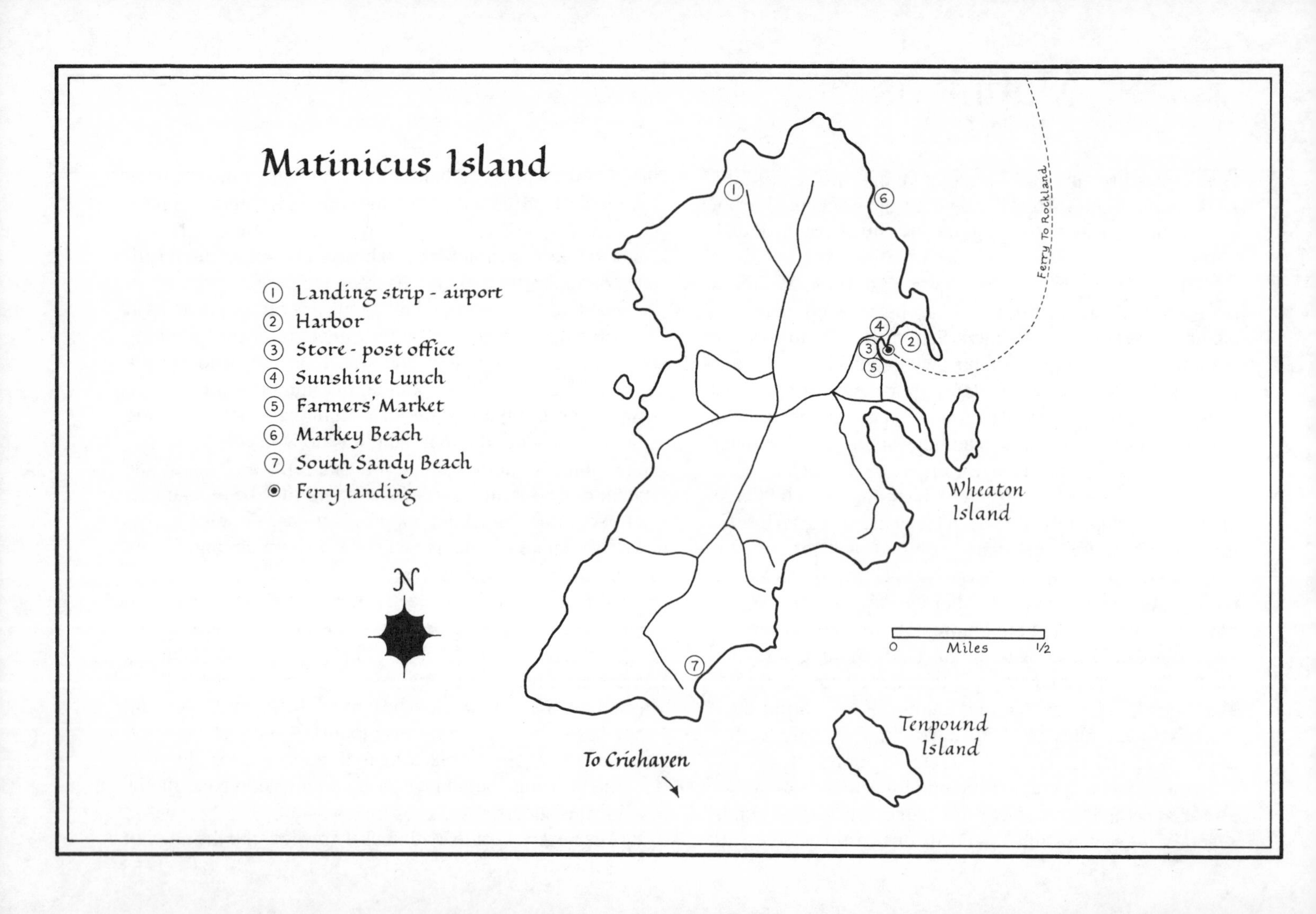
Matinicus Island
1 Landing strip - airport
2 Harbor
3 Store - post office
4 Sunshine Lunch
5 Farmers' Market
6 Markey Beach
7 South Sandy Beach
Ferry landing
N
Ferry To Rockland
Wheaton Island
0 Miles 1/2
Tenpound Island
To Criehaven

seat of the Cessna U206 six-seater. Pilot Charlie Jones sat next to me while the other seats held a Matinicus family and a lobsterman who had gone off-island for a bit of freshwater fishing.

To keep within the plane's 1600 pound limit, each of us was allowed only one bag, and these were packed in with the mail, a box of diapers, and the island Avon supply.

The twelve-minute ride was far too short, with panoramic views stretching to Mount Washington, 100 miles away, Cadillac Mountain on Mount Desert Island to the eastward, and Pemaquid Point to the west.

The islands below us looked like spruce-green rugs on the deep blue ocean surface—Pickering, Little Green, Large Green, Metinic—and before we knew it, the cluster of islands that includes Matinicus, Ragged (often called Criehaven), Wooden Ball, and Matinicus Rock. We landed on the gravel strip, and I headed toward town on foot until I was offered a ride by the family that had shared the plane. The airport is less than a mile from the center of town. From there, it is another half mile down to the harbor.

The Harbor

Matinicus harbor is one of the most picturesque on the coast, with its outer harbor crowded with lobster boats, and an inner harbor formed by the granite pier of the steamboat wharf (ferry landing).

Homes and fish houses, weathered in shades of grey, huddle in a group at the end of the harbor, each supported by long-legged stilts because of the 12-foot tides. Between and behind these buildings is a maze of boardwalks leading to wharves piled high with lobster traps, brightly colored pot buoys, and tangles of rope.

A young lobsterman, in his twenties, was painting the bottom of his boat as it lay on the mud of the inner harbor. He talked about coming out here as a sternman (helper) seven years ago. He worked for a lobsterman for a few years, then decided to settle here, and now has his own boat and home. But he has no wife as yet. "It takes a special kind of woman to live here," he says.

Matinicus women speak about the long, isolated winters, and how you "get through them." One says she sews, knits, joins friends for morning coffee and Jane Fonda exercise classes, and goes to meetings of the Ladies' Aid. "You have to keep busy," she says. "If you don't, you'll go mad."

I walked along South Sandy Beach and looked at the beach peas, goldenrod, Queen Anne's Lace, and raspberry bushes that grow at the edge of the sand. Sunbathing is good here, but the water is too cold for all but the most hardy swimmers.

Lobstering starts at an early age on Matinicus. Jeremy Van Dyne, 11, says, "I've been lobstering most of my life, since I was three." Jeremy and his friend, Steven Ames, 13, lobster together in Steven's open skiff with Jeremy's 15-horse outboard motor. I perched in the bow of their skiff, trying to keep my sneakers out of the brown, murky water that sloshed around in the bottom of the boat, and the boys took me out for an afternoon while they hauled a few of their 70 traps.

Steven sat atop the motor, steering between his legs, standing up to haul in the traps. Jeremy unloaded the lobsters — some three-pounders — measured them, threw out the crabs, fish, seaweed, and old bait, then rebaited the bags with herring. The boys grinned in anticipation as they opened each trap, as if it were Christmas morning.

We motored from cove to cove, a flock of sea gulls swarming after us and swooping down to snatch up bits of old bait as Jeremy tossed it out. The orange, glowing light of the setting sun sparkled against the fleet of lobster boats that lay at anchor as we returned at last to the harbor.

History

The island historian is Charles A. E. Long, who wrote *Matinicus Isle, Its Story and Its People*, giving detailed accounts of such items as the 44 wrecks, the flora, and the physical make-up of the island.

Indian people long used the island as a summer camp on which they gathered eggs, hunted sea birds, and fished. Vikings may have landed here, but the first written record is in John Josselyn's journal, describing the island in 1671 as "well supplied with homes, cattle, arable land and marshes."

The first permanent settlers were Ebenezer Hall and his wife and children, who came from Portland in 1751. Hall angered the Indian people by burning over the fields on nearby Green Island. Four Indians signed a letter of protest and sent it to the governor, but nothing was done.

In June 1757, a group of Indians landed on Matinicus, killed Ebenezer Hall, and kidnapped his wife and daughters. His son Daniel escaped out the window of the house, hid in the bushes until the Indians left, then hailed a passing boat for help. Another son, young Ebenezer, was out fishing at the time, and so escaped. A plaque on a rock down by Matinicus harbor tells today's visitors about the massacre.

Weathered gray cottages and fish houses line the shorefront of Matinicus harbor.

After the end of the French and Indian Wars in 1763, young Ebenezer Hall returned to Matinicus with his wife, whose maiden name was Young. This couple was soon followed by relatives, and many of today's islanders trace ancestry back to the Halls and the Youngs.

During the nineteenth century, Matinicus boomed as a fishing island, with the year-round population reaching 300. The school enrolled 61 scholars in 1869. For comparison, there were 22 students in 1925, and 9 students in 1983 in the one-room school (kindergarten through eighth grade). As with most of the smaller islands in Maine, the high-school age students must go to mainland schools and live with a family there, or else attend a boarding school.

How to Get There

Maine State Ferry Service, 517A Main Street, Rockland, ME 04841, (207) 594-5543. Service is provided one day per month. Contact the office to find out when a boat is scheduled for Matinicus.

Stonington Flying Service, Ash Point Road, Owls Head, ME 04854, (207) 596-6211, runs charter service to Matinicus, and takes passengers on the mail plane every weekday morning for $15.

Albert Bunker, Matinicus, ME 04851, (207) 366-3737 runs a charter boat service to and from the mainland. He carries the mail several times a week during foggy spells, and will take passengers along. He also runs charters from Matinicus out to Matinicus Rock, a five-mile trip.

Lodging

The island has no inn or bed and breakfast. Some islanders take people for a night or two, on occasion, and some people rent out their cottages when they are not in use. Send a letter to Postmaster, Matinicus Island, ME 04851, or telephone (207) 366-3755 to arrange a visit. Or contact Cait Bunker, Tourist Bureau, Matinicus Island, ME 04851, (207) 366-3443. Geoffrey Katz, 156 Francestown Road, New Boston, NH 03070, (603) 487-3819 rents his cottages when he's not using them.

Facilities for Yachtsmen

Matinicus has no harbormaster, no yacht club, and no official guest moorings. Local fishermen will be able to say if a mooring might be available for a night. Most yachts anchor in the middle of the harbor between Wheaton Island and the mainland. A dinghy may be beached or tied to the town wharf (there is no float).

Diesel fuel and gasoline are available at the town wharf, which has good water at half-tide or higher. Water can be hand pumped at the town well behind the store, but there is no hose on the dock.

Meals

Sunshine Lunch is a lunch wagon that serves hot dogs, hamburgers, seafood rolls, and drinks. People congregate here at lunch time, sitting at picnic tables with a great view of the harbor. On a rainy day there is no shelter. The hours are 11:00 a.m. to 3:00 p.m.

The store at the harbor sells ice cream, some snack food, staples, fresh fruits, vegetables, meat, beer, and wine. The Farmers' Market, run by Betsy Burr and Cait Bunker, is open 11:00 a.m. to 1:00 p.m. each Monday and Friday during the summer. The market offers fresh vegetables, fruits, breads, baked goods, and homemade jellies. Cait Bunker's goat's-milk fudge and goat cheese come from Matinicus goats. Plans are underway to sell island-made crafts in future.

What to Do

Exploring the island on foot is a favorite pastime, walking leisurely along the dirt roads or along the beaches (Markey Beach, South Sandy Beach, Condon Cove, or the beaches on the western shore). Swimming is a bit on the chilly side, but sunbathing is wonderful on the beaches.

Birding is excellent, according to bird lovers, any time from mid-May through the end of September. The migration in spring usually peaks the third week in May, and on a good day a birder may see as many as 75 species. The fall migration stretches from end of August through September. Good places for birding are the airstrip (horned larks), pond near the village store (warblers, green herons), Harbor Point, and the cemetery. Seabirds include double-crested cormorants, common eiders, gulls, terns, guillemots, ospreys, common loons, and many others.

Raspberries, strawberries, blackberries, and blueberries grow along the roadsides, but please don't go into people's back yards to pick.

Religious services are held in the church during July and August. During the remainder of the year, ministry is provided by the Maine Sea Coast Missionary Society via its boat, *Sunbeam*, which comes complete with minister as well as medical and educational services. Supported by many denominations and private individuals, the *Sunbeam* has traditionally served the off-shore island communities of Maine, visiting each once a month from October to May. For more information, contact Maine Sea Coast Mission, 127 West Street, Bar Harbor, ME 04653, (207) 288-5097.

A boat trip to Matinicus Rock, five miles southeast of Matinicus, can be arranged with Albert Bunker, who takes up to six people in his lobster boat. He usually circles the island and comes in close to the ledges where seals sun themselves. In very calm weather, Bunker can sometimes put people ashore.

Matinicus Rock, 32 acres of rock with nary a tree on it, has Maine's most famous seabird colony. People come long distances to view the thirty pairs of Atlantic puffins, with their large red beaks and webbed feet. The puffins arrive annually to dig their burrows and hatch their young. This is one of two long-standing Atlantic puffin colonies on the coast of Maine. The other is at Machias Seal Island near the Canadian border. A third colony has been recently established at Eastern Egg Rock in Muscongus Bay.

Other birds seen here include arctic and common terns, razorbills, shearwaters, eider ducks, and Leach's storm petrel. Suggested time to go is mid-June to mid-July, when nesting activity is at its peak.

The lighthouse was first established on Matinicus Rock in 1827. A story well known along the coast tells how Abbie Burgess, seventeen-year-old daughter of the lighthouse

Lobstermen paint boat during low tide in Matinicus' inner harbor.

keeper, kept the light going through a terrible storm in January 1856. Her father was stranded on Matinicus Island while getting supplies, and Abbie, there with her invalid mother and four younger sisters, spent a hungry three weeks waiting for the waves to calm down enough so that keeper Burgess could return.

On the way to Matinicus Rock, you will pass Ragged Island. The main harbor and village of this island are named Criehaven. Although once Criehaven was part of Matinicus Plantation, it is now a separate community, under state control. The houses are deserted in winter now, and there is no ferry service, store, or post office. The summer people who stay there use their own boats to reach the island. Criehaven is the "Bennett's Island" in Elizabeth Ogilvie's trilogy, *Storm Tide*, *Ebb Tide*, and *High Tide at Noon*.

Islesboro's Pendleton Point town park has rocky shores, pebble beaches, and berry bushes.

Islesboro

ISLESBORO IS TUCKED far into the north end of Penobscot Bay, where it is well sheltered from the rolling surge of the open ocean. It is eleven miles in length from Pendleton Point to Turtle Head, and its width varies from three miles at its widest to a few hundred yards at The Narrows. This 6,000-acre island offers many lovely views of Penobscot Bay.

The ferry ride from Lincolnville Beach is short and sweet, three miles in 20 minutes. I enjoyed the view from the open upper deck of the *Governor Muskie*. West Penobscot Bay was dotted with sailboats and fishing boats as well as larger vessels en route from Belfast to the ocean. The Camden Hills looked like a stage backdrop behind the salty scene.

An Englishman sitting next to me on the life-vest box was reciting poetry to a young lady. I was fascinated by his accent and stylishly casual attire. Thinking he must be a young British count or lord visiting Islesboro from abroad, I engaged him in conversation. He turned out to be a butler for one of the Dark Harbor "cottagers" The American young lady traveling with the butler was the newly hired downstairs maid. They were returning after a day off spent in nearby Camden.

The butler on the ferry was an indication of the wealthy and famous people who vacation in Islesboro. Visitors to this island might catch a glimpse of a celebrity: perhaps Walter Cronkite coming in off his yacht, members of the Kennedy clan, or novelist Sidney Sheldon, whose *Master of the Game* focuses on this island.

Dark Harbor is one of America's most exclusive summer resorts, dating back to the late nineteenth century. This area on the southern third of Islesboro includes a small village and about fifty estates. The "cottages" are more like mansions, some with households of up to 25 servants, including butlers, maids, nurses, chauffeurs, and boatmen. A few of these homes can be seen from the public road that runs down to the public park at Pendleton Point, but the best views are from the perspective of the yachtsman.

The ferry lands at Grindle Point next to a picturesque lighthouse. The keeper's cottage has been turned into a museum for local history. Unlike many island ferry landings, which drop passengers at villages, Islesboro's landing is out on a remote point, so visitors are advised to bring an automobile or bicycle.

There is no central village on this spread-out island. The post office, commercial garage, and a grocery store are in one location; the historical society and library are in another. School and town offices are in a third place. North Islesboro has its own store. The biggest and most attractive center is Dark Harbor, with a fine restaurant, two lunch counters, a gift shop, a laundromat, and a real estate broker's office.

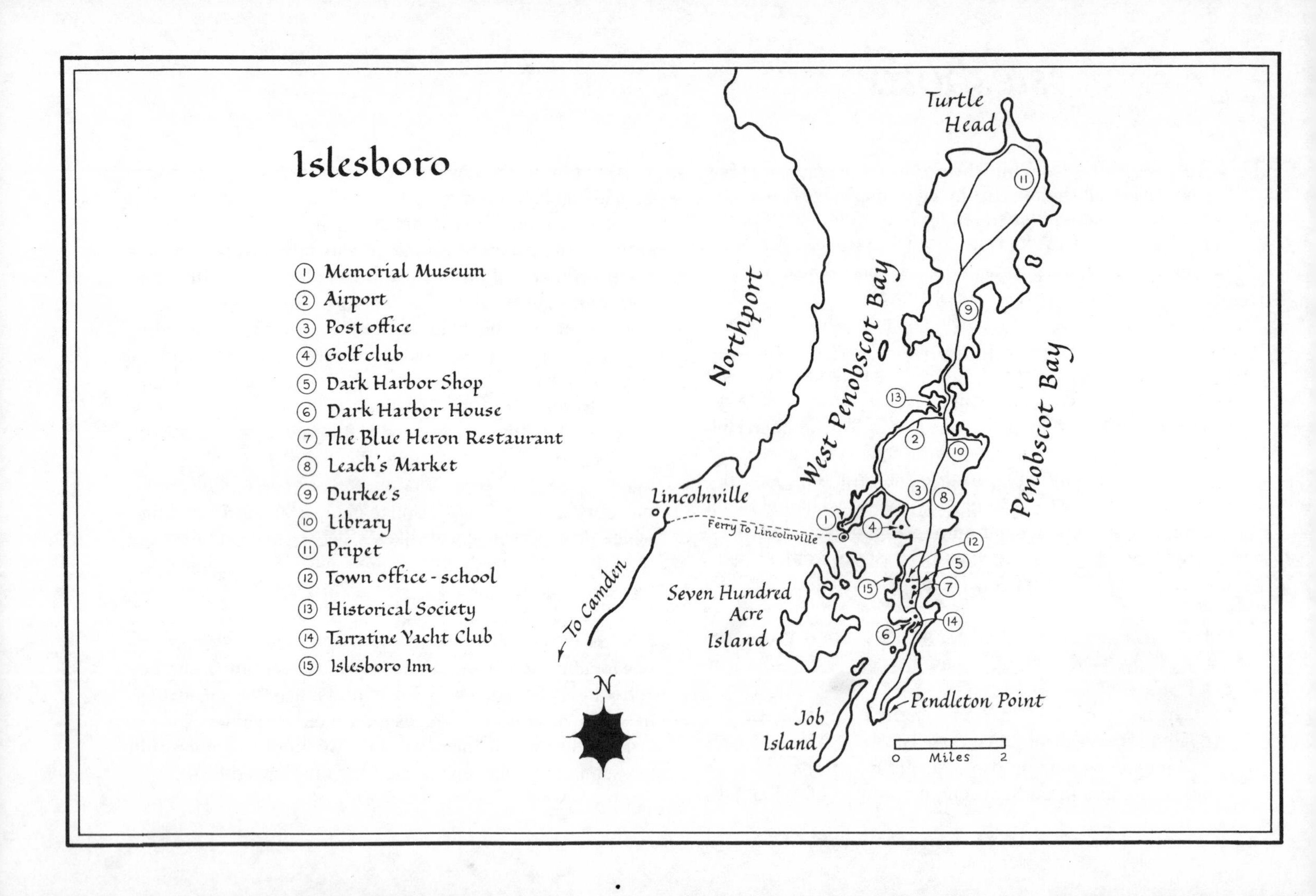
Islesboro
① Memorial Museum
② Airport
③ Post office
④ Golf club
⑤ Dark Harbor Shop
⑥ Dark Harbor House
⑦ The Blue Heron Restaurant
⑧ Leach's Market
⑨ Durkee's
⑩ Library
⑪ Pripet
⑫ Town office - school
⑬ Historical Society
⑭ Tarratine Yacht Club
⑮ Islesboro Inn
Turtle Head
Northport
West Penobscot Bay
Penobscot Bay
Lincolnville
Ferry To Lincolnville
To Camden
Seven Hundred Acre Island
Job Island
Pendleton Point
N
0 Miles 2

Bill Warren owns the Dark Harbor Shop—a bookshop, gift store, and lunch counter rolled into one. After making a lobster-salad lunch for a customer, Warren took me into the back room where he has his realtor's office. He showed me his listings, which ranged from $25,000 up to $550,000. Housing is scarce and many people rent summer homes for part of the season. A Dark Harbor Cottage, fully furnished with linens and silver service, could rent for as much as $12,000 a month. Realtors often inquire how much silver service a potential customer might require.

Now a year-round resident, Warren first came to the island during his childhood summers. His father and grandfather summered here. He comments, "Wild dogs couldn't send me back to Connecticut." He notes that the slow pace of island life was great while his children were young, but it's boring to teenagers. Islesboro's school, with about 90 students, covers kindergarten through twelfth grade, but often there are only two or three in a graduating class. Warren's 15-year-old son goes away to boarding school.

In contrast to many island communities, Islesboro's population is growing. According to Greg Marquise, the island's only resident attorney in 1983, the population has increased by a third in the last ten years. Year-round population is about 525, a figure that at least triples during the summer.

Proximity to the mainland makes living on Islesboro less arduous than living on a remote island such as Monhegan or Isle au Haut. Electricity comes by cable from the mainland. Automobiles can be transported across to Lincolnville with relative ease and low expense. Although there was no doctor as of 1983, a town-employed physician's assistant is on call to handle everything from minor cases to serious emergencies. Serious health crises often involve a special ferry trip to a mainland hospital.

Marquise remarks, "Economically, this island is very strong. Islesboro has had an infusion of new wealth." Island employment is mainly in the trades, such as carpentry, plumbing, or gardening. Wealthy summer people and retirees bring jobs and money to the island.

An increasing number of people find themselves in a position to live on Islesboro because they work by terminal plugged into a city computer. While this phenomenon is true of all Maine islands to some extent, Islesboro offers the added advantage of a location handy to the mainland when a business trip is required.

History

Called Long Island on a 1691 map of the New World, Islesboro was first settled by Europeans during the 1760s. At the time of the Revolution, when the British controlled this area of Penobscot Bay, 24 families lived on the island.

Despite the war, Harvard scientists obtained permission from the British command at Castine to observe the solar eclipse from Islesboro on October 27, 1780. Dorothy Simpson describes the episode in her book, *The Maine Islands*.

A small sailboat carrying colonial soldiers and a Harvard man sailed out to the British warship *Albany* off Castine and presented a letter of proposal: "Our business is solely to promote the interest of Science, which is the common interest of all mankind." The request was granted, with the instructions that the scientists not communicate with the island residents. A group of 10 scientists from Harvard—professors, graduates and undergraduates—anchored in Williams Cove off Islesboro.

In spite of the British restrictions, the scientists needed a base of operation, and Islesboro resident Shubael Williams secretly allowed them to use his home. After an anxious week of fog, the scientists successfully observed the eclipse under clear skies.

The township was incorporated in 1789 as "Islesborough," later shortened to its present spelling. For a hundred years, islanders fished and farmed for a living. All this changed with the advent of wealthy summer rusticators during the 1880s. Jeffrey Brackett, a young Harvard student from Boston, spotted this island as his steamer passed it on the way to the fashionable resort of Bar Harbor. He was one of many young people dissatisfied with the staid Bar Harbor atmosphere who were seeking something different. Brackett purchased 200 acres on Islesboro and described the shore to some of his Harvard friends who lived in Philadelphia. A group of them organized the Islesboro Land and Improvement Company of Philadelphia and purchased the southern third of the island—some 2,000 acres—for $100,000 in 1889.

They named the area "Dark Harbor" after the narrow inlet, not easily seen from the water, using "dark" in the sense of "obscure." The very next year, the organization built the elegant Islesboro Inn to house members and guests in 120 lavish rooms. The inn burned down in 1914 and another on the same site was torn down in 1953. Today's Islesboro Inn is situated on the western shore.

Tarratine Yacht Club is an active sailing and racing center at Dark Harbor.

Mrs. E. A. Daniels, who summered here for many years, wrote a description of the early times at Dark Harbor in her 1935 book, *Facts and Fancies and Repetitions about Dark Harbor by one of the Very Oldest Cottagers.*

"In those early years we got on without the game of bridge. We had euchre, sailing, dancing, lunches and dinners, golf, tennis and walking," she writes. "Yes, we did wear comfortable little woolen dresses of a morning . . . We had no telephones and no ugly telephone poles to mar our lovely views. . . ."

Mrs. Daniels lists other "amusements" such as bathing, riding, driving, and shooting seals. She says that on rainy days "some of the unregenerate" played poker in the boat house, but "lady guests wouldn't stand for it in the hotel." Among the distinguished guests of the Islesboro Inn were Mr. and Mrs. Charles Dana Gibson, Lord and Lady Astor, novelist Winston Churchill, and President Theodore Roosevelt.

James Murray Howe, one of the community's founders, described how people were chosen for the Dark Harbor set. "There was no special attempt to get Chicago people to the island," he explained. "In fact, Philadelphia and Boston citizens and those of good social standing were what we encouraged—on the whole I think with some success."

One of the largest estates was that of Charles Dana Gibson, encompassing almost all of nearby Seven Hundred Acre Island. The artist summered here from 1904 to 1944 during the period when he created his famous Gibson Girl drawings for magazine covers. His model was his beautiful wife. This island estate has its own chapel, hand-built by Gibson, and a miniature castle, complete with dungeons, that Gibson constructed to amuse his grandchildren. While a tourist or visiting yachtsman may land at Dark Harbor Boat Yard on Seven Hundred Acre Island, most of the rest of the island is the private residence of Gibson's descendants.

Automobiles were legally outlawed on Islesboro in 1913 as rusticators attempted to keep the island as a quiet retreat. Cars were not allowed until the year-round residents voted to change the law during the 1930s. The first cars came by scow in 1932 and a car ferry was established by 1936.

How to Get There

The Islesboro ferry leaves from Lincolnville Beach, located on U.S. Route 1 about five miles north of Camden. The Maine State Ferry Service boat, the 24-vehicle *Governor Muskie*, makes the three-mile trip to Islesboro in 20 minutes. Nine daily round trips during the summer make this the most accessible ferry-served, residential island east of Portland. Over 100,000 people and 50,000 vehicles make the trip each year, accounting for close to half of the total ferry use for the six islands under Maine State Ferry Service.

A summer day-tourist may board the first ferry in Lincolnville Beach at 8:00 a.m. and return, after nearly nine hours, on the last boat at 4:30 p.m., arriving back on the mainland by 5:00 p.m. People who bring vehicles might consider coming off island before the final, crowded ferry. On Sunday the early boat leaves Lincolnville at 9:00 a.m. and the last trip is same as weekdays. Off season, there are five trips daily and between three and five trips on weekends.

Most of the cost of the ferry service is subsidized by the State of Maine. Cost-saving efforts do result in periodic schedule changes, and it is advisable to double-check current timetables. The Lincolnville Beach terminal can be reached at (207) 789-5611. Address inquiries to Maine State Ferry Service, Lincolnville, ME 04849.

Islesboro's airstrip is to be paved in 1984 to better accommodate the hundreds of charter and private flights each year. Two commercial airlines handle most charters: Ace Aviation, Inc., Belfast Airport, Belfast, ME, (207) 338-2970, and Stonington Flying Service, Ash Point Road, Owls Head, ME 04854, (207) 596-6211.

Lodging

The Islesboro Inn, situated in a 16-room former private mansion on Gilkey Harbor, faces spectacular sunsets over the bay and the Camden Hills. The cocktail lounge and restaurant are open to the public. Activities include tennis, shuffleboard, croquet, Ping-pong, and walking through formal gardens and woods paths. Bicycles, small boats, and dinghies may be rented. Inquire about yacht charter trips. Make reservations through Kathleen Waterman, Islesboro Inn, Dark Harbor, ME 04848, (207) 734-2222.

Dark Harbor House offers bed and breakfast in a turn-of-the-century summer cottage built originally for one of the "Philadelphia 400." In spite of the elegance of the colonnade entrance and sweeping lawn, the inn's atmosphere is informal and friendly. The rooms are spartan — no T.V. — and bathrooms are shared. These bathrooms are old-world gems, with marble washbowls, tubs on claw feet, and chain-pull toilets. Although the inn has no water view, guests can easily walk the short distance to Dark Harbor or the yacht club. Dark Harbor House is open June through September. For reservations, write Dark Harbor House, Dark Harbor, Islesboro, ME 04848, (207) 734-6669.

The Town Manager sometimes has names of private homes where visitors may stay. Contact the Town Office, Islesboro, ME 04848, (207) 734-6445.

Since there are no licensed campgrounds on Islesboro proper, normal camping activity is effectively prohibited. The town of Islesboro donated nearby Warren Island to the State of Maine for camping purposes. Warren Island is used primarily by boaters because there is no ferry to it. For more information, check with Camden Hills State Park, Belfast Road, Camden, ME 04843, (207) 236-3109.

Facilities for Yachtsmen

Tarratine Yacht Club is a private club located on Ames Cove within walking distance of the village of Dark Harbor. The club dining room and porch overlook the anchorage. Tarratine Yacht Club, Dark Harbor, Islesboro, ME 04848, (207) 734-6994 (sailing master) or (207) 734-2281 (kitchen).

Pendleton Yacht Yard, situated just across the cove from the yacht club, has several moorings available for visiting yachtsmen. On occasion, the yard has small boats to rent as well. The dock has eight feet of water at high tide but no water at low tide. Gasoline, diesel fuel, and outboard mix are available. This yard hauls and stores yachts up to 40 feet in length. Pendleton Yacht Yard, Islesboro, ME 04848, (207) 734-6728.

Dark Harbor Boat Yard, situated in quiet Cradle Cove off

Bikers on Islesboro take a break to enjoy the view of East Penobscot Bay.

Lone yacht anchored in Islesboro's Crow Cove.

Seven Hundred Acre Island, has about 15 moorings available for rent during the summer. Known for its fine wooden sailboats, this yard produced the racing fleet known as the "Dark Harbor 20s." The yard can do mechanical repairs and keeps an inventory of Evinrude and Mako parts. The float has seven feet of water at low tide, and a water hose and fuel pumps (diesel, gasoline, and outboard mix are available). The yard carries block ice as well. Yachtsmen can take showers and do laundry here. Transportation is provided across to the main island of Islesboro. Dark Harbor Boat Yard, Box 196, Islesboro, ME 04848, (207) 734-2246. VHF channel 16 is continuously monitored by the yard.

Islesboro Inn has four moorings in quiet Gilkey Harbor and a dinghy float at the inn's dock. Yachtsmen may use a coin-operated washer and dryer in the inn basement. For a small fee, showers are available in "captain's cottage" near the shore.

Meals

The Blue Heron in Dark Harbor offers fine dining in an attractive setting, with white tablecloths, candles, and fresh flowers. Open for lunch and dinner daily except Monday. For reservations, phone the Blue Heron Restaurant, (207) 734-6611.

Islesboro Inn offers tea, cocktails, and/or dinner with saltwater view. (207) 734-2222.

Dark Harbor Shop serves sandwiches, pockets, quiche, salad, and ice cream cones at an inside counter or at outside picnic tables. Open daily 8:00 a.m. to 7:00 p.m.; Sundays 9:30 a.m. to 7:00 p.m. (207) 734-8878.

The Island Pub in Dark Harbor serves pizza and sandwiches as well as beer and wine and often provides live entertainment two or three nights a week.

Leach's Village Market in Dark Harbor has a tank with live lobsters that can be cooked to order, lobster and crabmeat rolls, hamburgers, french fries, and drinks. Open 8:00 a.m. to 5:00 p.m., summer only, (207) 734-6938.

Leach's Market, a grocery store, is located in the middle of the island near the post office. (207) 734-6672.

Mary's Kitchen in Durkees, a grocery on the north end of the island, has a deli and lunch bar serving home-cooked food, sandwiches, and pizza. (207) 734-2201.

The Snack Shack near the ferry landing offers crabmeat and lobster rolls, hamburgers, hot dogs, french fries, drinks and soft ice cream.

What to Do

Bicycling on Islesboro is delightful, with over 20 miles of paved roads. This long, narrow island affords numerous views of the water, looking east across Penobscot Bay to Cape Rosier and west across the water to the Camden Hills.

Pendleton Point on the southern tip is a good place for a picnic lunch. Areas thick with raspberry and blackberry bushes invite one to stop and pick on the way down to the public pebble beaches on several different coves.

Turtle Head Cove on the northwest end of the island has a very narrow strip of sandy beach. Turtle Head, the tip of the island, is privately owned land, but the owners, Mr. and Mrs. W. Henry Hatch, usually grant requests to walk there.

The Tarratine Golf Club offers a nine-hole course, open to the public for a fee. Call (207) 734-2248.

The *Flying Fish*, a 45-foot main topsail schooner, takes up to six passengers for day, evening, or overnight sails in Penobscot Bay. Meals are included. Contact Captain Earl MacKenzie, Islesboro, ME 04848, (207) 734-6714 or 734-6984.

The Alice L. Pendleton Memorial Library is open year-round. Summer hours: Wednesday, Saturday, Sunday, from 1:30 p.m. to 4:30 p.m., and Wednesday nights from 7:00 p.m. to 9:00 p.m.

Sailors' Memorial Museum, located on Grindle Point next to the lighthouse, has exhibits of Islesboro and vicinity maritime history. This museum is open 10:00 a.m. to 4:00 p.m., Tuesday through Sunday, summer only. Admission is free.

An excellent reference map of Islesboro is printed annually by the town and distributed, beginning around Memorial Day, at the ferry ticket office at Lincolnville Beach. A free copy may be obtained from the Islesboro Town Manager, Town Office, Islesboro, ME 04848, (207) 734-6445.

Lobster traps on the grass outside many Vinalhaven homes testify to the fact that this is a working fishing community.

Vinalhaven

VINALHAVEN HAS THE unspoiled charm of a fishing community, the shops and restaurants of a popular tourist town, and a fascinating history related to lobster, fish, and granite. Located fifteen miles east of Rockland in the middle of Penobscot Bay, Vinalhaven is the largest Maine island not connected to the mainland by a bridge. With an area of roughly ten thousand acres, the island measures seven miles by five miles. There are so many bays and inlets that no point on this island is more than a mile from salt water.

Most of the island is wooded with spruce or other coniferous trees and the soil is generally thin, spread on a rocky base. The terrain is fairly flat with a few hills; the highest peak is at Fox Rocks (215 feet above sea level) in the northwestern part of the island. Round Pond provides drinking water to the village.

One hundred years ago Vinalhaven was the center of a highly productive quarry industry, providing building blocks for such prominent public buildings as the Boston Museum of Fine Arts, Brooklyn Bridge, Grant's Tomb, the Cincinnati Post Office, and the Pilgrim's Monument at Plymouth, as well as paving stones for many city streets. At the height of the granite boom in 1880, Vinalhaven's population was 2,855.

Today's year-round population of about 1,200 relies on fishing as the primary economic base, supplemented by income from summer people and tourists. Most of the year-round islanders live in or near Carver's Harbor, the only town. Summer homes dot the rest of the island. Of the 39 miles of public roads, a good portion is paved.

The northern part, along the Fox Island Thorofare, is socially closer to North Haven than to Carver's Harbor. Summer people who live by the Thorofare traditionally use small motorboats to speed over to North Haven for a loaf of bread or to pick up their mail.

Vinalhaven has more modern conveniences than many islands along this coast: electricity from the mainland by cable, telephone service since 1898, and town water in the village. Community services include elementary and high school facilities, several churches, police and fire departments, a full-time dentist, and a medical center with resident doctor.

The Maine State Ferry Service runs a car ferry, *Governor Curtis*, between Rockland and Vinalhaven seven days a week all year long. The 13-mile trip takes one hour and twenty minutes. In all kinds of weather, it can be a beautiful trip. Sometimes a thin layer of fog turns bell buoys and fishing boats into shifting ghostly shapes. Other times, when the air is crisp and clear, every detail of the seascape is super sharp.

The Fox Islands
(Vinalhaven and North Haven Islands)
North Haven Island
① Ferry landing
② North Haven Vil.
③ Boat yard
④ Almon Ames' Inn
⑤ Bulli Ruffian Inn
⑥ Mullen's Head Park
⑦ Scenic vista
⑧ Ames Knob
⑨ Church
⑩ Landing
⑪ Golf course
⑫ Scenic view
⑬ Private airstrips
Vinalhaven Island
① Booth's Quarry
② Isle Au Haut Mt.
③ Arey's Neck Woods
④ Brown's Head Light
⑤ Tip Toe Mt.
⑥ Private airstrip
⑦ Lawson's Quarry
⑧ Geary Beach
North Haven Island
Webster Head
Pulpit Harbor
Ferry To Rockland
Southern Harbor
Fox Islands
Thorofare
Perry Creek
Zeke Pt.
Calderwood Neck
Seal Cove
Winter Harbor
Seal Bay
See Opposite
0 Miles 1
Vinalhaven Island
N

Vinalhaven Village

1. Main Street
2. Ferry landing
3. Armbrust Hill
4. Grimes Park
5. Tidewater Motel
6. Library
7. Bridgeside Inn
8. Sands Cove
9. Lane's Island Preserve
10. Historical Society
11. Galamander Exhibit

Approaching Vinalhaven, the ferry ducks into a narrow passage behind a row of rocky islands, each neatly capped with a shock of blue-green spruce trees. As the captain weaves the cumbersome craft through a long, narrow passage between Greens Island and Vinalhaven, he maneuvers a slalom course of red "nun" and black "can" buoys that mark the channel. The ferry docks at the wharf in Carver's Harbor.

On the next wharf, lobstermen crate their wiggly, green catch, soon to be stuffed and baked for customers in restaurants in Boston, New York, or perhaps San Francisco. I met a twelve-year-old Vinalhaven boy who spoke with the confidence of an adult about the business of lobstering. Already in his third year as a lobsterman, he owned his own skiff and fifty traps.

One old timer spoke about Vinalhaven life: "I wouldn't live anywhere else. Being 'round the water is wonderful. People are friendly. I know most of the faces even if I can't put names to them. Nothing is rushin'. You can take time to do what you want to do." He added that there is plenty to do on the island for a person who has the energy. "It's amazing how busy you can be, but I don't thrash to it now."

The village that curves around the harbor is a tourist's delight, with many small gift shops, lunch counters, restaurants, bike and moped rentals, and the only motel on any off-shore Maine island. Even an apartment complex sits on the hill, a low-income building for the elderly.

Storekeepers are friendly. At the hardware store, chairs invite people to sit and chat. Proprietor Bruce Grindle, now in his eighties, greets everyone who comes in. "Have a

Open air restaurant at Vinalhaven's Sands Cove serves lobsters and seafood, picnic style.

candy," he told me, offering a cast-iron frying pan full of goodies. For children, Grindle has special delights: a bag full of small raisin boxes, apples, and cookies. His scrapbooks of photographs are wonderful. He has a photo of every Vinalhaven resident, including all the children, each photographed sitting in the hardware-store chair with the same glass case in the background.

Day tourists in the summer have five hours between boats, from about 10:00 a.m. until 3:00 p.m., to explore the island. A favorite pastime is to take a picnic to one of the public nature preserves. Lane's Island, owned by the Nature Conservancy, is close to town yet has a good view out to sea. The trail around the perimeter takes about half an hour, depending upon how long one spends picking raspberries.

Vinalhaven has blueberries and blackberries, but the raspberries are best of all. Our family had a summer home on Vinalhaven when my children were young. The raspberries were so thick that we made them into pancakes, muffins, and jam, yet never seriously dented the supply for the birds.

Vinalhaven is one of the few Maine islands with freshwater swimming. The abandoned quarries now filled with rainwater are delightful. In places the granite is cut into broad steps, providing warm rocks for sunbathing and shallow pools for young children. The very clean water is warm, in contrast to the cold Maine ocean. Some quarries have steep rock sides that drop deep into the water from great heights. From these sheer cliffs, I have watched Vinalhaven boys perform swan dives, dropping down like terns diving after fish.

History

Martin Pring is credited with discovering Vinalhaven and North Haven in 1603 and with naming this group of islands the Fox Islands after observing a number of silver-grey foxes there. At that time these islands were inhabited by Indians who successfully discouraged settlement by white people for another hundred and fifty years.

The first permanent settler was David Wooster in 1760. He was joined by others after the end of the French and Indian Wars.

The Fox Islands were treated as a unit during the 1700s. They were incorporated as one township under the name Vinalhaven in 1789. William Vinal was one of the first assessors (selectmen) and a prominent citizen. His father, John Vinal of Boston, was the legal agent who arranged the incorporation.

Vinalhaven at that time encompassed quite a number of islands, including what is now North Haven, Hurricane Island, and Matinicus. Matinicus was "set off" in 1840, and North Haven became an independent town in 1846. Hurricane Island was incorporated in 1878 at the peak of the granite boom when this tiny island had a population of 600 people. In 1937, after the quarry business was abandoned, Hurricane again came under Vinalhaven's administration.

Since Penobscot is the largest bay along the Maine coast, it became the natural border between French Acadia and the British possessions. Settlers on Penobscot Bay islands found themselves in the midst of a struggle for many years. In the early days, it was the French and Indians against the British. During the Revolution, most Fox Islanders sided with the rebels, but the British used these islands to house troops and also confiscated food and livestock. Island men were forced to leave their families and work on building a British fort at Castine. Again during the War of 1812, British ships captured coastal schooners and disrupted trade.

Weather has also caused tragedies. One oft-repeated tale is the story of the *Royal Tar*, a steamer en route from Yarmouth to Boston in the fall of 1836. Caught in a northwest storm, the ship sought shelter near Diamond Rock, under the lee of Vinalhaven. Aboard the boat were 20 crew members, 72 passengers, and a group of circus animals, including camels, lions, horses, a leopard, a tiger, a gnu, and an elephant. The boat caught fire and a few crew members and passengers left in the lifeboats, leaving many people and the animals aboard the flaming vessel. The animals were forced overboard to sink or swim. Captain Dyer of North Haven saw the fire and took his vessel close enough to save forty passengers. Historians are uncertain whether the survivors were taken to North Haven or to Isle au Haut. Drowning victims included 29 passengers, three crew members, and presumably all the animals. The carcass of the elephant washed up on nearby Brimstone Island.

When the quarry business began to boom in the 1880s, as many as 1,500 men were employed. Italians, Swedes, Welshmen, Scots and Englishmen worked in the quarries by day and drank and sang the nights away.

The largest company, Bodwell Granite, began in 1871 and operated until 1919. In 1899, Bodwell was commissioned to quarry and lathe four 64-foot columns for the Cathedral of St. John the Divine in New York City. Each stone, measuring 64 feet by 8 feet by 7 feet, weighed over 300 tons, the largest stones ever removed from any quarry. One can imagine the heartbreak when the first three monoliths cracked under the vibration of the lathe. To complete the job, new stones were quarried and each column was pieced together in two sections. The massive monoliths were transported to New York, where they now stand in the cathedral.

By the 1930s, cement began to take the place of granite and the island granite quarry business ended. Vinalhaven men turned once more to the sea to earn their living.

Lobstering has been important since the mid 1800s. Vinalhaven had several lobster canning plants, the first one started in 1846. The Boston firm of Johnson and Young closed off a saltwater basin near Carver's Harbor in 1884, forming a pound for keeping lobsters alive until winter when prices go up. Over a hundred thousand lobsters were held in this natural lobster tank.

Fishing has also been important. During the late 1700s and early 1800s, a salt works on Calderwood Neck provided the salt for curing the millions of pounds of fish caught each year. Boiling 400 gallons of sea water would produce one bushel of salt.

In 1852, John Carver developed a business of knitting horse nets (to keep flies off the animals' backs) and selling the nets on the mainland. By 1900 a thousand people were employed on Vinalhaven making horse nets.

How to Get There

The *Governor Curtis*, operated by the Maine State Ferry Service, is the ferry between Rockland and Vinalhaven. The 15-mile trip takes an hour and 25 minutes. The boat makes three trips a day, Monday through Saturday, during the summer season. The first trip leaves Rockland at 8:40 a.m. and arrives in Vinalhaven just after ten o'clock. The last boat leaves the island at 3:05 p.m. On Sundays the early boat leaves Rockland at 8:45 a.m. and the last boat leaves Vinalhaven at 3:30 p.m. In winter the boat goes twice a day during the week and once a day on Sundays.

The *Governor Curtis* carries 17 vehicles, but tourists are advised to think carefully about taking a car to the island, particularly during August weekends. In order to get a place on the return ferry home, a driver may have to put the car in line and move it up with every ferry all day long. Reservations are recommended. Day visitors do not need a car because the ferry docks conveniently close to the village, near enough to walk to most points of interest.

Lodging

The Tidewater Motel, once a blacksmith shop for the bustling quarries, is located in the center of the village. The rooms are attractively furnished and have private baths. Some units have sundecks built over the harbor; some units have kitchenettes. The Tidewater is open all year. It serves no meals but is located near several restaurants. Contact Arthur and Pat Crossman, Vinalhaven, ME 04863, (207) 863-4618.

The Bridgeside Inn is located on Indian Creek, ¼ mile from the village. No meals are served; guests often buy take-out meals in town and bring them back to the dining porch overlooking the water. The rooms are decorated in old-fashioned island style with flowered wallpaper, braided rugs, and paintings by local artists. Guests in the nine rooms share two and a half baths. This quiet location has a view of the water and is close to Lane's Island Nature Preserve. The inn is open July and August only. Contact Peter Goodwin, (summer) Bridgeside Inn, Main Street, Vinalhaven, ME 04863, (207) 863-4854; (winter) 420 Woodlawn, Philadelphia, PA 19144, (215) 849-4427.

Peterson's Camps has four cabins outside of town, with kitchens and outside plumbing. Cabins sleep two to seven people and are fully furnished with dishes, sheets, and blankets. The camps are rented by the week or longer. Contact Winona Peterson, Vinalhaven, ME 04863, (207) 863-4836.

Facilities for Yachtsmen

Vinalhaven has no yacht club. When the Army Corps of Engineers dredged Carver's Harbor, the town was required to provide three guest moorings, which should be available by summer of 1984. For permission to use one of the moorings, contact Harbormaster Harold Chilles, Vinalhaven, ME 04863, (207) 863-2216. Chilles is a fisherman, and, when he is on his boat, he can be reached on Channel 6, VHF, by asking for Harbormaster Vinalhaven.

There are three floats in town. One is at Calderwood's wharf, where gasoline, diesel fuel, oil, water, block ice, and groceries are available. This is a busy dock, so do not stay here any longer than necessary to get supplies.

Vinalhaven granite quarries, no longer used by industry, have become popular swimming holes.

The town float is at the head of the harbor. The town also owns the fish company float. Dinghies may be tied at any one of these three floats.

Yachtsmen wishing to anchor are encouraged to ask the harbormaster for advice.

Meals

Sands Cove Lobster and Clambake is situated directly on the shore, with a view of the lobster boats in quiet Sands Cove. Meals are served on picnic tables outdoors (no inside seating). Boiled lobster, steamed clams, crabs, and steak are served, along with homemade rolls, salad, and dessert. Meals are offered Wednesday, Saturday, and Sunday, from noon to 6:00 p.m., weather permitting. Special lobster bakes may be arranged for private groups at other times. The season is Memorial Day to Labor Day. Sands Cove Lobster & Clambake, Vinalhaven, ME 04863.

The Haven is open in the summer only. Breakfast and lunch are served Tuesday through Saturday; dinner is by reservation only on Tuesday, Wednesday, Friday, and Saturday. Call (207) 863-4969.

The Harbor Gawker is a take-out window with a tank where you can pick your own live lobster. Picnic-basket dinners feature lobster, haddock, or chicken. Beer and soft ice cream are also available. While there are no tables, meals can be taken next door to the Millrace, run by the same owner. Both places are seasonal. (207) 863-9365.

The Millrace, in the village, has a pool table and coin-operated games in a large open room where teenagers congregate in the evening. Open for breakfast from 4:30 a.m. until 11:00 a.m. Reopens at 4:00 p.m., with pizzas available.

The Pizza Pit has pizza, sandwiches, wine, and beer. Open mid-May to late October. (207) 863-4311.

Tibb's, a tiny lunchroom on Main Street, is open for breakfast, lunch, and dinner, hours: 4:00 a.m. to 8:00 p.m. (207) 863-2544.

The Gooey Wagon, located near ferry landing, serves hot dogs, hamburgers, sandwiches, and soft drinks (summer only).

Burger-Ped offers short-order meals to go: thick crabmeat rolls, hamburgers, and milk shakes. It's located opposite Calderwood's Wharf Island Grocery, between the ferry landing and village center.

There are also several grocery stores on the island.

What to Do

Vinalhaven has a number of parks within walking distance of the village. Grimes Park, next to the State Ferry Terminal, is two and a half acres of woodland with two small beaches, good views of the sea, and picnic tables. The granite watering trough in the park was used to water the oxen and horses that hauled stones from the quarries.

Armbrust Hill, a wooded park with trails, is situated ¼ mile from the last stores in the village, just behind the Island Community Medical Center. On this hill William Kittredge started Vinalhaven's first commercial building stone quarry in 1840. James Armbrust purchased the hill around 1870 to cut paving stones for mainland cities. This area has four quarries and several smaller pits, called "motions."

Trails are maintained on Armbrust Hill and have been planted with flowering trees and shrubs such as azalea and

Typical lobster boat of the Maine islands has a house to protect lobsterman from wind and rain. The after deck is large enough to carry stacks of cumbersome traps.

rhododendron bushes, which flower in the spring, and mountain ash, which has colorful fall berries. From the top of the hill it is possible to see Matinicus and several other islands in the bay. A children's playground with swings and slides is located on the section of Armbrust Hill just behind the medical center.

Lane's Island has a 45-acre nature preserve located a half mile from town, over a bridge. The surf on the bold outer shore can be spectacular on windy days. This land was given to The Nature Conservancy to be used in perpetuity as a

Lane's Island nature preserve, one of many preserves owned and managed by the Nature Conservancy, offers the visitor a pleasant place to walk along the shore.

recreational preserve. Visitors are requested to take away their trash. Do not light any fires, hunt, or camp here. This preserve is open at all times without admission charge.

On a rainy day, visit the Carnegie Library or Historical Society. The library is open from 2:00 p.m.until 4:30 p.m. and again from 6:30 p.m. until 9:00 p.m., Monday through Saturday. The Historical Society exhibits include quarrying tools, quilts, and many interesting old photographs. The Historical Society building is a former church, brought from the mainland by barge over 80 years ago. It is open from 11:00 a.m. until 3:00 p.m.

An outdoor exhibition of an authentic galamander, a wagon used to haul the huge granite blocks from the quarry to the wharf, is on permanent display in a park near the east end of the village. The wheels of the galamander are over six feet tall.

The main street has several gift shops, including the Island Gift Shop on Main Street, the What-Not Shop, the Crafty Cat, and the Laughing Pirate. The Old Drugstore boasts an antique soda fountain, used as decoration, and a gallery in back for displaying work of local artists.

Two abandoned granite quarries now filled with water provide good swimming holes. Lawson's Quarry is about a mile along the North Haven road to the north. Booth Quarry, stocked with trout, is east of town about the same distance and not far from Narrows Park on the ocean. Swimmers are requested not to use any soap or shampoo in order to keep the water clear and pure. While these quarries are open for public swimming, don't expect life guards, changing houses, or bathrooms.

To explore further afield, visitors might want to rent bicycles or mopeds, available at Burger-Ped, not far from the ferry landing. A fleet of 10 mopeds and 4 bicycles can be rented during June, July, and August. A bike route map of the island is sold at the Paper Store. There is no taxi.

Arey's Neck Woods has a trail leading to the rocky shore of Arey's Cove. This whole area along Geary Beach is good for beachcombing.

Browns Head Light Station, at the northwest end of the island on the Fox Islands Thorofare, is one of the few lighthouses along this coast still manned by the Coast Guard. From the lighthouse one can look across the water to North Haven and beyond to the Camden Hills.

Tip Toe Mountain, on the same road as Browns Head Light Station, is really a steep outcrop of volcanic rock. This town park has picnic tables, situated in a birch grove next to the huge rock. From the top there is an excellent view of Crockett's Cove and Dogfish Island.

Catholic, Episcopalian, Protestant, Reformed Latter Day Saints, and Christian Science churches have weekly services.

Vinalhaven has more arts and entertainment than most off-shore residential islands in Maine. Community concerts feature internationally known musicians. First-run movies are occasionally shown between June 1 and mid-October. A three-day arts and crafts show in August attracts many visitors to the island.

Windjammers off North Haven in the Fox Island Thorofare.

North Haven

NORTH HAVEN IS ONE of Maine's finest yachting centers. Situated on the Fox Island Thorofare, the short-cut through Penobscot Bay, North Haven watches the world go by. On a breezy summer's day sleek yawls tack through the narrow waterway, their sails full and their hulls heeled over. Not only yachts, but trawlers, draggers, sardine carriers, and lobster boats ply this route.

On summer Saturdays, the Thorofare is dotted with the white sails of North Haven dinghies. These small wooden boats originated on North Haven and represent the oldest racing class of sailboats in America.

Windjammer schooners give the Thorofare a nostalgic look. Formerly cargo vessels in the coastal trade, these schooners now take vacationers on week-long sailing trips in Penobscot Bay and surrounding waters. While home ports for these schooners are on the mainland at Camden, Rockland, Rockport, or Belfast, North Haven is a favorite stopover.

Each year, on the eve of the Annual Great Schooner Race from North Haven to Rockland, some fifteen of these vessels sail in at sunset to anchor off North Haven, filling the Thorofare with the sounds of a great party as passengers and crews sing on deck and shout back and forth between the boats.

Aboard the ferry from Rockland, the *North Haven*, passengers can watch all the activity of the Thorofare. The ferry chugs through the busy waterway with the deliberate air of a Mack truck in a crowded city street before it docks at the tiny village on North Haven.

The Island

The island of North Haven is eight miles long and three miles wide, encompassing 5,281 acres. The shoreline —nearly 50 miles in length—is full of coves and harbors, including beautiful Pulpit Harbor. North Haven has a few hills and its high point is 160 feet above sea level. A freshwater pond in the middle of the island provides drinking water for the town as well as a skating rink in winter.

In the early nineteenth century, when the island was farmed extensively, there were many open fields. Much of this land has now grown in with spruce and birch as the predominant trees. Still open are a few of those lovely yellow hay fields leading down to the water.

Most of the roads run inland from harbor to harbor rather than along the shore, where large tracts of property are privately owned. The year-round population tends to cluster in or near the village.

The village includes a grocery store, two gift shops, a knitwear shop, a lunch take-out window, a fish market, the library, the post office, a boat yard, and a school (kindergarten through twelfth grade).

North Haven's summer community includes people

who live on both sides of the Thorofare and use outboard motorboats or old-time launches the way mainlanders use cars. Large, comfortable "cottages" line the shores, many of them built in the 19th century when the first "rusticators" and their servants came by steamer from Boston. Generation after generation, these families continue to summer here, so that today there are great clans of Cabots, Saltonstalls, and Lamonts.

Ever since summer people came in the 1880s, this island has developed a tradition of offering privacy and anonymity to the rich and famous. Islanders treat each person according to how he, or she, acts on North Haven rather than by any position that person might have elsewhere. Charles and Anne Morrow Lindbergh often visited Anne's family on the Morrow estate. These days, Thomas Watson of IBM arrives by air, landing on his private strip near his summer home.

United States presidents have visited this island. In 1873 President Ulysses S. Grant stayed at the Mullen House. This inn predates the summer people. It was started by Mr. Mullen, the customs agent at North Haven when this town was a port of entry. Next to the Customs House, Mullen had a store as well as the inn. Another president who loved North Haven was Franklin Delano Roosevelt. In 1933 he anchored in Pulpit Harbor on the *Amberjack* and signed the register at the Mullen House.

In his book, *Better Than Dying*, North Haven islander William R. Hopkins relates another visit. "I think I have told you that in 1943 when Roosevelt and Churchill met in mid-ocean to sign the Atlantic Charter, Mr. Roosevelt, his energies spent, retired on the same USN destroyer which took him to his mid-ocean meeting, to a five-day hide-a-way at Pulpit Harbor . . . During that visit, the wire services were desperately trying to discover the whereabouts of the President of the United States. The UPI, acting on a hunch, telephoned North Haven, which, in those days before the dial system, had a 'central' operator. Orrie Woodworth was on duty when they called, and when asked if the President was at Pulpit Harbor, replied, quite accurately, 'I ain't seen 'im.'"

About four hundred people live on North Haven year-round, and the population increases in summer to about twelve hundred. In the 19th and early 20th centuries, a sharp social line divided the natives and the summer people. First Selectman Lewis Haskell recalls growing up on a summer estate where his parents were employed full-time. "It was an 'Upstairs, Downstairs' situation," says Haskell. This kind of year-round employment is no longer the norm. Islanders are more apt to work for summer people on short-term specific jobs, such as remodeling a house.

In recent years, the social division line has grown fuzzy. Many islanders in their twenties and thirties seek jobs elsewhere, returning to the island in later years. They may have college educations and hold excellent positions on the mainland. Some people who grew up summering on these lovely islands have decided to stay year-round, making a living with lobstering, scalloping, or doing whatever jobs they can find.

See Map On Page 72

Marion Hopkins is one of many local women who knit sweaters for the North Haven Yarn Shop.

There is less steady work available these days, not enough to employ all those who wish to live on the island. North Haven Future Group, including both summer and year-round residents, is brainstorming this issue, investigating how to expand the island economy without destroying the essential character of the island. This issue is not unique to North Haven, but is a central issue on nearly every island along the Maine coast.

Some of the islanders have unusual talents. Dorothy Brown carves beautiful birds out of wood, displayed in a window at J.O. Brown and Son, the boatyard next to the ferry landing.

Another artist, Eric Hopkins, combines two careers, fishing and art. His blown glass has been displayed in shows and museums in several major cities. A native islander, Hopkins uses forms of fish, shells, and other sea life as a theme. Hopkins lives in Rhode Island in winter, returning to the life of a North Haven fisherman each summer. "It's important to get away and then to come back in order to see things," says Hopkins.

Chellie Pingree buys up fleece from the Cheviot sheep on North Haven's Sky Farm to use for a knitting business. Fifteen or twenty island women knit for Pingree's yarn shop, making custom sweaters, hats, and mittens. Another North Haven woman, Jane Parsons, hand-spins her own yarn from these sheep and is teaching others the ancient art of spinning.

History

North Haven is the site of the oldest items found on any Maine island. An archeological dig on the site of an ancient

North Haven's main street has gift shops, a take-out sandwich bar, a grocery store, a boat yard, the ferry landing, the post office, and the library.

cooking pit on the Turner Farm came up with a specimen dated before 3000 B.C. The ancient Down East tradition of the clambake may be responsible for this find; the calcium content in shells neutralizes the corrosive elements in the soil. Many artifacts and bone fragments have been found around ancient island shell heaps.

Indian people succeeded in keeping white settlers off North Haven for a hundred and fifty years after Martin Pring discovered and named the Fox Islands in 1603. During this period, Indians attacked fishermen who came near the islands. Dummer's War (1722–25) was started by one such incident, related in the *The North Island: Early Times to Yesterday*, by Norwood Beveridge.

On June 14, 1722, a group of Indians attacked Jacob and Daniel Tilton in the Fox Island Thorofare and tied them up. Daniel tricked the Indians into untying him, saying he would find something to give to his captors. Once free, Daniel cut Jacob's bonds and the two of them overpowered the Indians and sailed away.

Beveridge relates another unusual story about Indians in the North Haven area. Abenaki Indians massed a huge fleet of 50 canoes in 1724. They attacked English vessels in the Fox Island Thorofare, killed about half of the Englishmen, and impressed the rest as sailors. Some Micmac Indians joined the group. Before long, the Indians had a fleet of 22 vessels, including a large schooner armed with a pair of swivel guns. The Indian fleet headed into the St. George River but was unsuccessful in trying to take the fort there.

The first white settler, David Wooster, came to the North Island (now North Haven) in 1762. Shortly after the Treaty of Paris (1763) many settlers began to come to the Fox Islands. Not only was the French and Indian War officially over, but there was an additional reason. A royal order of 1763 barred migration of unauthorized settlers to the land beyond the Alleghenies, so pioneering types from Massachusetts headed north and east along the coast of Maine.

So many people came to the North Island that there were nearly 800 by the mid-nineteenth century. In 1846, the North Island was set off from Vinalhaven and the township of North Haven was officially incorporated. North Haven people combined fishing and farming for their living. Another pursuit was boatbuilding, not only fishing boats for local use, but also large ocean-going vessels as well. The granite-quarry boom never hit North Haven

Soon after the first summer rusticators arrived on the North Haven scene, J.O. Brown's boatshop on the Thoroughfare was established in 1888. The boatyard is still operating, nearly one hundred years later, under the ownership of J.O. Brown's grandson, Jim Brown. Another boatworks was in operation at Pulpit Harbor by 1890.

How to Get There

The Maine State Ferry Service runs a car ferry, the *North Haven*, between Rockland and North Haven. The 12½-mile trip takes one hour and ten minutes. From the end of May to early September the boat makes three daily round trips, Monday through Saturday, and two round trips on Sunday. A day tourist may catch the 9:20 a.m. boat out of Rockland, arriving North Haven at 10:30 a.m., and leave on the afternoon ferry at 3:00 p.m. allowing for 4½ hours to explore the island.

On Sundays, the last boat leaves the island at 4:20 p.m., giving a visitor an extra hour and twenty minutes. Off-season, there are two trips daily and one trip on Sunday. For more information, contact Maine State Ferry Service, P.O. Box 645, 517A Main Street, Rockland, ME 04841, (207) 594-5543.

Travel between Vinalhaven and North Haven is not easily arranged for tourists. Most people plan to go to one or the other of these two islands. There is an informal inter-island service between the village on North Haven and a small ferry landing on the Vinalhaven side of the Fox Island Thorofare about seven miles from Carver's Harbor. J.O. Brown boatyard in North Haven will provide a boat to take passengers across the Thorofare. Phone ahead to J. O. Brown & Son, Inc., North Haven, ME. 04853, (207) 867-4621.

Air transportation by small charter plane from Owls Head, near Rockland, to North Haven is available through Stonington Flying Service, Ash Point Road, Owls Head, ME 04854, (207) 596-6211.

There is no official taxi service but innkeepers are helpful about arranging transportation for their guests.

Lodging

Almon H. Ames runs a small inn furnished in the comfortable turn-of-the-century style of many island homes. There are four rooms for guests, with one shared bath. A sundeck overlooks Brown's Cove. Ames does much of the cooking himself and enjoys putting on gourmet dinners for his guests. Open year-round, this inn is located in the village. Make reservations with Almon H. Ames, North Haven, ME 04853, (207) 867-4853.

The Bulli Ruffian is another small inn, particularly catering to yachtsmen who would like to come ashore for a day or two to stretch their legs, have a bath, and do laundry. There is room for about eight people and stays are limited to a few days. The rooms are tastefully decorated, as is the dining room. Animals complete the country atmosphere with horses grazing in the pasture, chickens, geese, seven cats, and a basset hound. This inn is located near the summer airstrip, but is not very convenient to the village (about two miles away). The inn is open from early April to the end of October. For reservations contact Henree Comley, Crabtree Point Road, North Haven, ME 04853, (207) 867-2219.

Haskell Village at Bartlett's Harbor has rental camps on a weekly basis. Reservations are requested; call (207) 867-4635.

Mabelle Crockett rents rooms in her home and will provide meals. Contact her at (207) 867-4708.

Mullen Head Park has a camping area with outhouses, and a pump for drinking water. Prior reservations must be made at the Town Office, North Haven, ME 04853, (207) 867-4433

Facilities for Yachtsmen

Only a few of the islands discussed in this book have yacht clubs. North Haven's yacht club, the Casino, is one of the oldest in Maine. Situated on the Fox Island Thorofare in the village, the club has a wharf and several floats and maintains two guest moorings. There is a simple clubhouse on the wharf, but no elaborate dining facilities.

The town maintains guest moorings as well. Yachtsmen interested in using a town mooring should contact the harbormaster, Foy Brown, through J. O. Brown & Son, Inc. Fuel, ice, water, and marine supplies are also available at the boatyard. The outer float has about five feet at dead low water. Brown's yard builds boats and makes repairs on boats and motors.

Meals

Bulli Ruffian serves candlelight dinners, offering one main course each night at a set price. The attractive dining room can seat 36 people at small tables, but only by reservation. Tea is served on the brick patio on Tuesday through Saturday. Open April through October. Contact Henree Comley, (207) 867-2219.

The Landing, a lunch window in the village, serves sandwiches, hot dogs, hamburgers, crab rolls, soda, and ice cream cones. Patrons sit on benches on a wide deck. Open summer and early fall.

Waterman's Store, a grocery, has lunch fixings. Open year round, Monday through Saturday from 7:00 a.m. until 5:00 p.m.

What to Do

The village shops along the waterfront have high-quality items. Hopkins Store has books, pottery, handknit sweaters, and an art gallery. The North Haven Yarn Shop offers handmade knitwear made by North Haven women during the winter. Those with the North Haven Yarn label are knit with wool from North Haven sheep.

The library, located on the main street, has an interesting collection of Indian artifacts and arrowheads. Library hours are 10:00 a.m. to noon and 3:00 p.m. to 5:00 p.m. daily in summer and fall; once a week in winter.

The historical society, located in a room off the Town Office, has books and old photographs on North Haven. Open Thursday, 2:00 p.m. to 4:00 p.m. during July and August, and by appointment.

Several boatyards include J.O. Brown and Son, Inc., next to the ferry landing, Edwin Thayer's Y-Knot Enterprises where boats are stored for the winter, and Douglas Stone's boatbuilding shop.

North Haven Rentals, Inc., has bicycles for rent. This fleet of 20 new three-speed Ross bicycles is the best group of rental bicycles I came across on any Maine island. Boats may be rented as well, but only a few are available. Contact Almon Ames (207) 867-4853.

The nine-hole golf course is open to visitors for a small greens fee. Golf carts are available.

Ames Knob is about a mile from the village, an easy climb to a high rock with a good view of the Thorofare. On a clear day, one can see Blue Hill to the north and the Camden Hills to the west.

Mullen Head Park has hiking trails, picnic tables, and swimming in the chilly waters of East Penobscot Bay. The park is located about four miles from the village.

Abandoned house at Head Harbor on Isle au Haut.

Islands East

THE DRIVE EAST AROUND Penobscot Bay is well worth the time and effort. This is the country of the pointed firs, the land of blueberries and lobsters, where villages have old-fashioned country stores and white churches with steeples. Around every bend is a beautiful view, perhaps a cozy harbor of fishing boats or the broad expanse of Penobscot Bay sparkling with the sails of a hundred yachts.

As the gull flies, it's only 20 miles from Camden to Stonington, yet by car it is 85 slow miles — maybe two hours — to travel north around the large body of water that is Penobscot Bay. Mount Desert Island is even further along the road. Add to that the characteristic isolation of islands and you can begin to understand the comment of one Isle au Haut woman: "In the winter, you travel 45 minutes by boat to Stonington and you're nowhere."

On the eastern side of Penobscot Bay, five year-round island communities are served by regular ferry service. Isle au Haut has a passenger/mailboat service from Stonington, on Deer Isle. Swans Island is the only one of these islands that has a daily car ferry. Run by the Maine State Ferry Service, the boat connects Swans Island with Bass Harbor on Mount Desert Island. This same boat makes two trips a week to Frenchboro, the village on Long Island, but this is not enough to provide access for a day visitor. The Cranberry Islands — Great Cranberry and Islesford (Little Cranberry) — have daily transportation by commercial boat operating out of Northeast Harbor on Mount Desert Island.

Isle au Haut

ISLE AU HAUT IS ONE of the most beautiful islands off the coast of Maine. Blessed with several harbors and rugged cliffs that rival those of Monhegan, the island is six miles long and three miles wide, encompassing 5,500 acres, of which 2,860 is Acadia National Park land. Its five "mountains" do not compare with the Rockies, or even with Mount Washington in height, but nevertheless they look impressive from the deck of a small boat. The highest one is Mt. Champlain, 556 feet above sea level, by far the tallest peak on any offshore Maine island. A long, narrow inland lake, Long Pond, has water clean enough to drink. A little over a mile long and 54 feet deep, it has a small, somewhat sandy swimming beach at the southern end.

Acadia National Park

Most of the visitors to Isle au Haut head for the beautiful remote woodland managed by Acadia National Park. The eight-mile boat trip to Isle au Haut from Stoning-

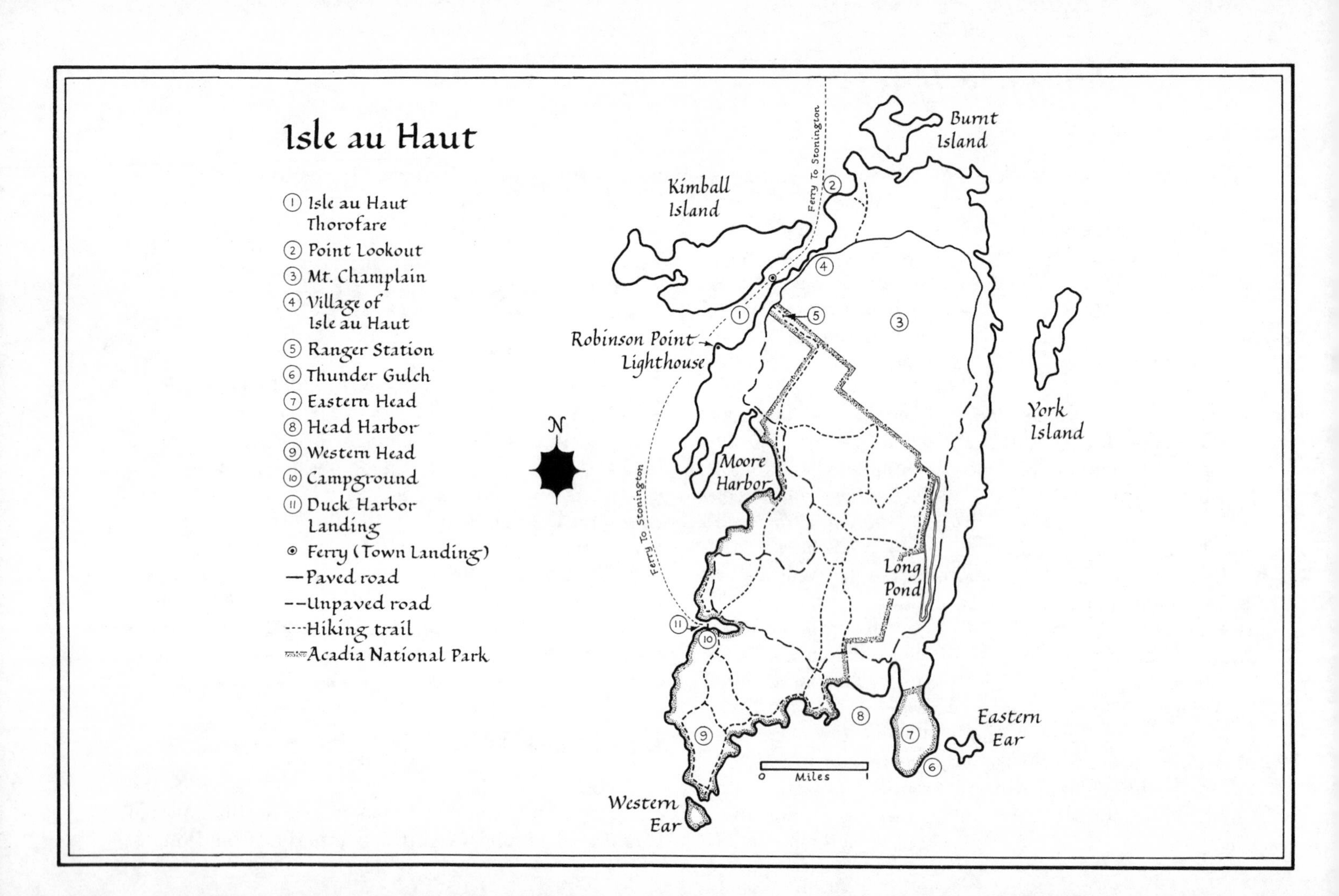
Isle au Haut
1 Isle au Haut Thorofare
2 Point Lookout
3 Mt. Champlain
4 Village of Isle au Haut
5 Ranger Station
6 Thunder Gulch
7 Eastern Head
8 Head Harbor
9 Western Head
10 Campground
11 Duck Harbor Landing
Ferry (Town Landing)
Paved road
Unpaved road
Hiking trail
Acadia National Park
N
Burnt Island
Kimball Island
Ferry To Stonington
Robinson Point Lighthouse
Moore Harbor
York Island
Long Pond
Eastern Ear
Western Ear
0 Miles 1

ton snakes through what some sailors consider to be the prettiest cruising ground in the world. Known as Merchants Row, this is an archipelago of small islands ringed with granite ledges and topped with thick stands of spruce and some balsam fir. The boat lands at the Duck Harbor on the southwest side of the island.

Forty-eight percent of Isle au Haut is part of Acadia National Park. The main part of this park is on Mount Desert Island, accessible by car. Of the approximately four million people who visited the park in 1983, about 3,000 came to the Isle au Haut section. To protect the fragile environment on Isle au Haut, Acadia National Park Authority has set a top limit of fifty visitors at any one time. Park land covers the southern and western shores as well as most of the central hills and woods. (Check your map or ask the ranger about boundaries of the park.)

The boat landing for Acadia National Park is situated on Duck Harbor, where day visitors and campers disembark. A park ranger usually meets the boat, passes out maps, and explains the trail system, giving an idea of how long each loop usually takes. Day visitors have six hours on the island.

I chose the 2½-mile loop around Western Head, hiking through a moss-carpeted forest out to the open rocky ledges. Cliff Trail and Goat Trail follow the southern shore, winding in and out of the trees, sometimes following cairns along the beach. Back in the woods I came across piles of gray, smooth beach stones, thrown there by fierce winter storms.

Hikers have many choices of trails. Most people choose the paths along the dramatic southern cliffs, from which one can look down at surf breaking below. Some folks walk along the road to town, where they catch the boat (at the town landing) on its way back from Duck Harbor to Stonington. A few ambitious hikers take Long Pond Trail over the mountains to the village. Unfortunately, there is no broad scenic view from the central ridge or the wooded top of Mt. Champlain. A better view can be had from Duck Harbor Mountain near the boat landing following a pleasant four-hour loop around Western Head.

The Village

The mailboat from Stonington lands at the village, on the northwest side of Isle au Haut. The town is on the Thorofare, open at both ends, which runs between Isle au Haut and Kimball Island. It's a pretty spot, with a lighthouse, a church spire, and often numerous fishing boats and yachts at anchor. The village is simplicity itself: a store/post office, church, town hall/library and school. No gift shop, no restaurant, no inn, no frills. In winter about 70 people call Isle au Haut home. In the summers the population rises to about 300.

Many of the summer folk live at Point Lookout, ¼-mile north of the village on the Thorofare. Dating back to the 1880s, Point Lookout has boardwalks, large rambling "cottages," and a landing for the passenger boat from Stonington.

Recently, the year-round residents and summer folk developed a common interest in protecting their island from the influence of the National Park. Many people feel threatened by the increasing numbers of park visitors. The social environment on small islands is extremely fragile and can change radically when overloaded with people. Until recent years, tourists landed at the village and hiked across the island to the park. A few visitors even treated the village as a living history museum for the park. Townspeople were pleased when a special boat route was established direct from Stonington to the Acadia National Park section of the island, bypassing the village. The relatively few people who hike back to the village are welcome as long as they respect private property.

Surprisingly few of the year-round residents are native to the island. Many newcomers have come seeking a lifestyle of simplicity and self-sufficiency. These folks are referred to as "new year-rounders," to differentiate them from "year-rounders."

A new year-rounder who has been part of the community for over eight years comments: "It's exciting and wonderful to live in a community in such a beautiful spot." He speaks of becoming more attuned to the cycles of nature, as in preparing for winter. "It's a challenge," he says. While surviving winter is difficult, this man claims the biggest challenge comes from social and cultural isolation "that wears most subtly."

This life of isolation is also a life fraught with danger. The sea is powerful, the water is icy cold and perilous. In May 1983, the night before Mother's Day, three Isle au Haut residents drowned in a tragic boating accident when their skiff filled up with water as they motored home from watching a movie in Stonington. Two survivors clung to the boat for seven hours until daylight when a helicopter picked them up.

Like many of the outer islands, Isle au Haut has little support from the mainland in the way of services. Islanders

Deer become tame on Isle au Haut where they are protected from hunters.

maintain the roads as best they can. Until recently most of the island's electricity was produced by a central generator. In 1983, 52 year-round and seasonal islanders shared the cost of laying an electric cable from the mainland. The island still has no telephone service, but most seasonal people don't want it; they use citizen's band radios instead.

The island residents cooperatively own the store and hire a manager to run it. Stock includes fresh meat, canned and frozen food, fresh produce, hardware, fuel oil, and gasoline for the island's unusual assortment of automobiles.

Since there is no car ferry, the only way to transport a vehicle to the island is by barge, at great expense. These vehicles are kept alive by any means possible through efforts that rival those of a doctor saving a dying patient. It is commonplace for a car to be started by touching two wires together under the hood. Gas might be hosed to the motor from a five-gallon tank in the back seat, as if the car were an outboard-motor-driven skiff on wheels. In its final stage, the island car is put out to pasture behind the house or along the road, where over the years it gradually becomes camouflaged by bushes.

There is but one basic road around the island, blacktopped through the town and part-way around the east side. The western part of this circle is more like a wide fire trail, usable only by heavy duty trucks, hikers, or joggers. I rode my bike along the east-side road, past the sawmill and Long Pond to Head Harbor, where there used to be an active community. Most of the homes are now used in summer only. Six yachts lay at anchor in this quiet harbor.

Through the woods on the eastern side of Eastern Head lies Thunder Gulch, a narrow crevice in the cliffs where spray may shoot fifty feet in the air during a storm. I walked out to the point along the ledges, climbing over large boulders along the way. From time to time I stopped to photograph a tiny flower growing out of a crack or to check out the sea life in a tide pool.

History

The first residents of Isle au Haut were Indians who came to the island in summer, perhaps driven to the islands by the insects inland. They dug clams and gathered the eggs of nesting birds.

This island was named in 1604 by Samuel de Champlain who called it "Isle Haute," later corrupted to "Isle au Haut." Some people give it a French flavor by pronouncing it "eel a ho," while fishermen and local islanders usually say "aisle a ho."

The first white settlers came to nearby Kimball Island in 1772, and an early survey shows five inhabitants on Isle au Haut by 1785. Peletiah, Henry, and William Barter settled on the Thorofare in 1792. By 1802 the community was established enough for 24 people to petition the Commonwealth of Massachusetts (which then included what is now the state of Maine) for title to the Isle au Haut land.

This fishing community grew steadily during the 1800s and Isle au Haut's population hit its high point of 274 residents. In 1850, several families built a schooner in the Thorofare and sailed off around Cape Horn to the gold fields, but local historians have no record of whether or not they made it.

Lobstering became the predominant occupation in the

late nineteenth century. In those days. Lobsters were so plentiful that they could be picked up in tide pools at low water or scooped up from the shallows with a small net. A Boston firm established a lobster cannery on Isle au Haut in 1860 and shipped to such clients as Crosse and Blackwell in London. After the cannery closed in 1873, lobsters were shipped to Boston and New York in boats that allowed sea water to circulate through a special section in the hold.

Ernest W. Bowditch of Boston visited the island in 1879 and told his friends about it. The next year, he and a few others formed Isle au Haut Company and purchased land at Point Lookout for a bachelor's retreat with no women, no children, and no dogs. As these men eventually married Point Lookout became a family summer resort. Bowditch acquired a great deal of land on Isle au Haut, and in 1945–46, his heirs donated most of the land to Acadia National Park to be kept as a wilderness accessible to the public.

How to Get There

Isle au Haut Company runs two separate trips to Isle au Haut: the mailboat to the village and the tour boat to Duck Harbor at Acadia National Park. The two excursion boats each hold about fifty passengers. The *Miss Lizzie* is named after Isle au Haut's beloved long-time postmistress, and the second vessel is the *Mink*. Trips are scheduled on mail days, Monday through Saturday, with only special charter trips available on Sundays and holidays.

The mailboat service goes three times a day to the village in summer, an eight-mile, 45-minute ride. Even in the off-season, the boat makes at least two round trips a day.

The Duck Harbor tour boat operates mid-June to late September, Monday through Saturday, with charters on Sunday. It leaves Stonington at 11:00 a.m. and arrives at Duck Harbor at noon. The return trip leaves Duck Harbor at 6:00 p.m., arriving at Stonington by 7:15 p.m. No reservations may be made for either boat. During the summer season it is permissible to park a car in the Stonington school parking lot up the hill from the opera house. For more information, contact Isle au Haut Company, Stonington, ME 04681, or telephone (207) 367-5193, Monday through Friday, 8:00 a.m. to 5:00 p.m. or Saturday 8:00 a.m. until noon. Schedules are subject to change so it is wise to call ahead.

Lodging

Acadia National Park has five Adirondack-type lean-to shelters at Duck Harbor. Nicely situated in the woods, with a fair amount of privacy, each lean-to holds a maximum of six people and has a picnic table and fireplace next to it. Drinking water is available from a pump. A chemical toilet is located by the road; two compost toilets serve the campground. Dead wood may be gathered for the fireplaces. When I visited in August, campers said that the mosquitoes were still pretty bad at night. Tents are only allowed on the wooden platforms. Deer flies are sometimes a problem. It's a good idea to bring fly dope or bug spray and a large piece of netting to cover the mouth of the lean-to.

The campground is only a quarter mile from the Duck Harbor ferry landing. Before mid-June or after mid-September, when the Duck Harbor ferry is not in service, campers must backpack four miles from the village to the campsite, about a three-hour hike.

The Miss Lizzie *and the* Mink *transport mail and passengers from Stonington to the village and to the Acadia National Park section of Isle au Haut.*

Beachcombing on Isle au Haut's Acadia National Park.

Because of the island deer population, estimated by the park service to be between 300 and 500 animals, dogs are not allowed to stay in the campground, and may be taken hiking only if they are leashed at all times. Reservations are taken by mail after April 1, and often fill up completely by June, or even sooner. People are requested to reconfirm ten days before their reservation, and some may cancel at this time, creating unexpected openings. Before mid-June or after mid-September there may be vacancies. The campground is open from mid-May to mid-October. Camping outside of the designated shelters is strictly forbidden. If one misses the boat home, one should immediately contact the ranger. Make reservations or check on information through Acadia National Park, Box 177, Bar Harbor, ME 04609, (207) 288-3338.

Facilities for Yachtsmen

Isle au Haut has many beautiful harbors, including Head Harbor, along the dramatic southern coast; narrow Duck Harbor, where Acadia National Park has its boat landing; and Moore Harbor on the west side. Most picturesque of all is the Isle au Haut Thorofare, with the tiny town and white church steeple outlined against the dark hills.

The island has no yacht club and no guest moorings. Yachtsmen who anchor in the Thorofare and wish to explore the town may tie dinghies to the float at the town wharf. Just down the road is the store, which has its own high-water wharf (five feet of water at high tide). While the store sells diesel fuel and gasoline to fishermen, and can spare a few gallons for a yachtsman in trouble, the supply is not ample enough to fuel visiting boats. There is no water hose. Stonington, only ten miles away on the mainland, has such supplies.

Meals

Isle au Haut has no inn and no restaurant. The general store in the village has sandwich fixings and occasionally offers steamed hot dogs. Park visitors who intend to hike from Duck Harbor to the village and hope to buy a snack should first check the store hours with the ranger.

What to Do

Hiking on the trails is the usual activity in Acadia National Park, along with bird watching and photography. Bring rain gear and a warm sweater, and wear good walking shoes or hiking boots as trails may be wet and rough. Ask the ranger's advice on which trails to take. Hikers often spot deer that are relatively tame, for there is no deer-hunting season here. Fire is a terrible worry for people who love Isle au Haut, so smokers are urged to be very careful with cigarettes and obey no-smoking bans. In dry periods, smoking may be prohibited on trails and campers may have to use stoves rather than campfires.

Taking a bike to Isle au Haut can be disappointing because roads are mostly unpaved and have sections of soft sand. Ten-speed racing bicycles are particularly poor. Mountain-type bicycles with large tires are recommended.

Walking or jogging is preferable. For a beautiful all-day walk, follow the circular road around the island (about 11 miles), stopping at Long Pond for a swim and picnicking at one of the beaches.

Swimming is best at a small man-made beach on the south end of Long Pond. On the ocean beaches, stones have been rounded smooth by the wave action, but the water is very cold.

Help keep this beautiful island unspoiled by picking up trash—your own or any other trash that you see along the trails—and taking it off the island.

Fine Sand Beach on Swans Island lies alongside a spruce and fir forest, facing the waters of Toothacher Cove.

Swans Island

SWANS ISLAND LIES five miles southwest of Mount Desert Island. Although historians assert that this island was named for Colonel James Swan, who purchased it in 1786 with the intention of making a fortune, the name also suits the island's shape, which from above resembles a swan with great wings outstretched.

At its extremities, this 7,000-acre island measures about six miles across. The irregular shoreline winds in and out around numerous deep coves and pointed fingers of land. The Carrying Place in the middle of the island is so narrow that Indians used to portage canoes across it. Swans Island is moderately hilly, with two mountains: Goose Pond Mountain (240 feet) and Big Mountain (210 feet). The populated areas around Burnt Cove Harbor have open views of the water; undeveloped waterfront and inland sections are more apt to be heavily wooded.

Unlike most islands discussed in this book, Swans Island has three villages, each with its own post office: Atlantic has the ferry landing, library, and museums; Swans Island (village) is a fishing settlement on the west side of Burnt Coat Harbor; Minturn lies across that same harbor, with a restaurant and the swimming quarry. The three villages make up one township with one group of selectmen supervising everything from taxes to winter ice breaking in the harbor. Telephone service is by microwave radio relay.

The year-round population of roughly 350 people more than doubles in the summertime to 750. One elementary school, kindergarten through eighth grade, serves the whole island, with three teachers and 39 pupils in 1983–84. High school pupils commute by ferry to Mount Desert Island High School.

Although Swans Island is not as tourist oriented as Monhegan, it does offer paved roads for bicycles, beaches, a museum, at least two gift shops, and many beautiful vistas. My last visit was in September when the weather was clear and crisp. Burnt Coat Harbor waterfront buzzed with activity. Fishermen shoveled slippery silver herring into bait house bins, joking with one another in Down East style. One of the windjammer schooners sailed into the harbor in the late afternoon, and its passengers explored the village later that evening.

Swans Island is in transition, with no less than three residential developments underway in 1984. Such large-scale land development was not evident on any other island I visited.

The Island Retreat (270 acres) on Back Cove and Mackerel Cove, is the oldest housing development, started in the early 1970s. Most lots are sold and about a dozen homes have been built. Within this area are 110 acres of permanent park land. Red Point, along the eastern shore facing Mount Desert Island, is about 700 acres, 10 percent

Swans Island and Long Island

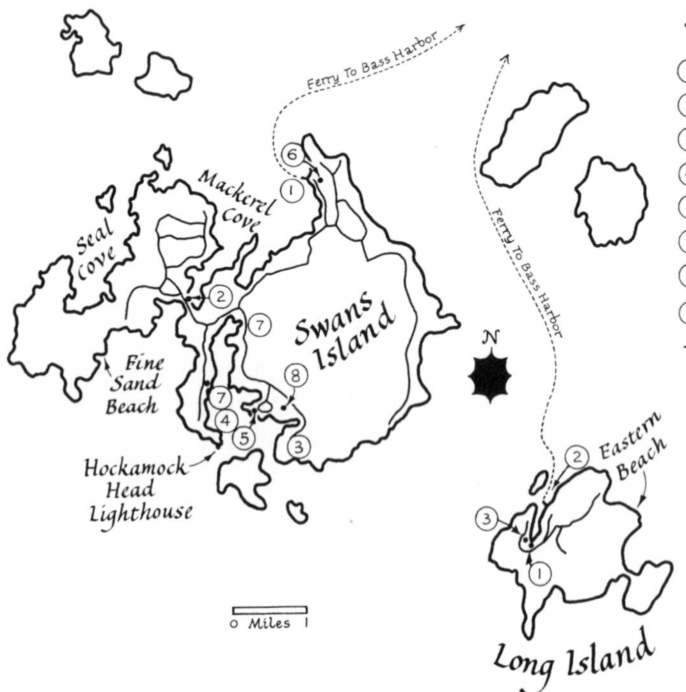

Swans Island

1. Ferry landing
2. The Carrying Place
3. Buswell's Cabins
4. Burnt Coat Harbor
5. Sea Breeze Restaurant
6. Museum
7. Grocery Stores
8. Quarry pond

— Road

Long Island

1. Frenchboro (village) church and school
2. Ferry landing
3. Historical Society

— Road

of the island. Only a small portion of the tract has been sold and about ten homes completed. West Point (99.5 acres) facing Isle au Haut, is still in the road-building stage.

Islanders have differing opinions on development. Alberta Buswell, the wife of a retired lobsterman, states, "We thought it would help on taxes, but I can't see that it is helping one bit. Our taxes go up anyway." She comments on the additional cars in the ferry line. "When we want to get off (the island) on the ferry, it's bad."

Some of the most vocal opponents of development are long-time summer residents. One who has lived in areas of Long Island and Delaware where development has changed things predicts the same for Swans Island: "In ten years, it won't be the same island." She hates to see large-scale residential development encroach upon her island refuge.

First Selectman Sonny Sprague sees new homes as a positive step, essential for economic survival. "If we didn't have the summer population, Swans Island couldn't survive as a year-round community," he says. Sprague points to the increased cost of public services. The Swans Island budget (including the three villages on the island) has grown from $30,000 in 1968 to $198,000 in 1983. He comments: "Developments don't cost us a thing except congestion on the ferry boat." Sprague thinks the type of people who have settled on Swans Island for vacation or retirement homes have been an asset to the island. "We have the finest summer people we could possibly have," he declares.

However, he is the first to admit that a section of Swans Island village between his home and Hockamock Head is beginning to look like "a ghost town in winter," with only a few year-round homes.

History

As on most of the Maine coastal islands, Indians summered here, leaving shell heaps from their clambakes. Prior to 1725, a year-round, coastal-dwelling Indian tribe, the Malecites, lived here, hunting seals and porpoises. Perry D. Westbrook, in his book *Biography of an Island*, mentions skeletons found at Burying Point near Hockamock Head that may have been Malecites killed by the plague of 1618–19.

Swans Island was once named Brule-Cote, coming from the French meaning burnt hill. Tradition holds that Champlain named this island after sighting a burned-over woodland here during his explorations around Mount Desert Island in 1604. Recent scholars dispute this, saying there is no mention of this island on Champlain's maps. They attribute the name to some unknown early French fisherman. Brule-Cote was anglicized to Burnt Coat Island.

The first white settler in this area was Thomas Kench, a soldier of English background. After surviving the grueling march to Quebec under Benedict Arnold in 1775, in which many soldiers died of starvation, bitter cold, or the smallpox epidemic, Kench returned to Boston. He volunteered again for a revolutionary artillery outfit led by Captain James Swan.

Soon after he volunteered, Kench deserted the army in 1776. He headed east along the coast and ended up in the Swans Island area, settling on Harbor Island at the mouth of Burnt Coat Harbor. There he lived the life of a hermit for over ten years.

Ironically, Kench's former officer, Colonel James Swan, purchased Swans Island and 24 smaller adjacent islands, including Harbor Island, in 1786. Swan was an entrepreneur who dreamed of making a fortune out of this island kingdom.

He had a gristmill and a sawmill constructed on Burnt Coat Harbor at what is now Minturn and built a mansion for himself on the land above. Swan offered 100 acres to any man who brought a family here, cleared the land, built a home, and stayed seven years. A number of people accepted this offer and moved their families to Swans Island.

Swan subsequently lost his fortune and escaped to France where he was imprisoned for a debt he claimed he did not owe. He remained in a Paris prison for 22 years until his release in 1830, just before he died.

Among the early settlers was David Smith, who arrived in 1791. It is Smith, not Swan, who is known as the first permanent inhabitant. He had such influence in the community that he was called "King David." Smith had three wives, at least 24 children, numerous grandchildren, and countless descendants, many of whom continue to reside on Swans Island. Smith's grave can be seen in the old cemetery on the north side of the island.

By 1834, after Swan's death in Paris, almost 200 people lived on Swans Island. Because of Swan's wheelings and dealings, land titles were in a legal mess. Despairing of dealing with bureaucracy, Swans Island residents gathered and formed their own town structure, taking squatters' rights ownership of their lands, making a tax list, and planning for construction of roads and a school.

Fishing has been the traditional economic base. During the 1870s, Swans Islanders were the most successful mackerel fishermen on the Maine coast. According to one account, 15,000 barrels of mackerel were caught by Swans Island men in 1879. When the mackerel began to disappear in the 1890s, the fishermen turned to herring and lobsters. Almost a hundred years later, lobstering is still the backbone of the Swans Island economy.

How to Get There

The *Everett Libby* leaves from Bass Harbor on Mount Desert Island for the six-mile, 40-minute trip to Atlantic, on Swans Island. The vessel carries both passengers and vehicles five or six trips daily (four on Sunday) in summer, mid-June to mid-September. The earliest boat leaves Bass Harbor at 7:45 a.m. (8:45 a.m. on Sundays) and the last return boat to Mount Desert Island leaves Swans Island at 4:30 p.m. The off-season schedule cuts back to three or four trips daily. No ferry runs on winter Sundays.

Since the ferry lands several miles from the main points of interest, it is advantageous to bring a car or bicycle. The problem lies in getting the car off the island. During busy summer weekends, one may have to wait for two (or even three) ferry circuits before getting a vehicle space on board.

A first-time visitor might want to purchase a reservation over and back. Three reservations are available for each trip, and they may be purchased at the ticket office a minimum of two hours before ferry time. Reservations must be made no more than 30 days from the date one wishes to travel. Contact Maine State Ferry Service, Bass Harbor, ME 04653, (207) 244-3254.

The island has no taxi in 1984. Bicycles are a popular means of tourist travel around this island. Bring your own bike, or rent one on Mount Desert Island at Southwest Cycle, Southwest Harbor, ME 04679, (207) 244-5856.

Kent's Wharf, one of two active lobster and fish wharves in the village of Swans Island on Burnt Coat Harbor.

Lodging

Alberta and Ted Buswell rent rooms in their home, a licensed facility overlooking Long Cove in Minturn. As I walked in, I could smell the sweet fragrance of apple pie cooking in the oven. Ted sat in the living room knitting lobster-trap heads as he watched television. This simple Maine-style home has housed celebrities such as members of the Kennedy family. The Buswells also accommodate the summer policeman who comes over for the season. Alberta has a friendly, forthright air and welcomes strangers to the island, saying, "If I don't have room in the house, I'll hang you in the closet."

Alberta Buswell also handles reservations for cabins on the water in Minturn. Contact Alberta Buswell, Minturn, ME 04659, (207) 526-4127.

People with vacation homes often rent them out to visitors for a week or two. Contact realtor Peg Bailey of the Knowles Company, The Carrying Place, Swans Island, ME 04685, (207) 526-4122.

Facilities for Yachtsmen

Swans Island has a number of popular harbors, including Mackerel Cove (where the ferry lands) on the north side of the island, Buckle Island Harbor and Seal Cove on the west side, and Burnt Coat Harbor on the south shore.

Gasoline and diesel fuel are available at the Fisherman's Co-op on the west side of Burnt Coat Harbor. Yachtsmen should request permission at the office before leaving a dinghy at the float.

Burns's Boatyard on Burnt Coat Harbor has a marine railway and storage facilities for yachts up to 32 feet long. The yard can do emergency repairs as well. Russell Burns, Swans Island, ME 04685, (207) 526-4135.

Marine mechanic George Tainter may be reached at Swans Island, ME 04685, (207) 526-4280.

Hindman's Harborside Cash Market in the village of Swans Island on Burnt Coat Harbor takes a special interest in yachtsmen. Hindman has plans to build a facility including showers, a laundromat, and a sandwich bar, to be in service by 1984 or 1985.

A boat launching ramp is available near the ferry landing in Mackerel Cove. There is also a dinghy float in this cove.

Meals

Sea Breeze Take Out in Minturn, overlooking the harbor, offers home cooking, fish dinners, and homemade desserts at reasonable prices. Open year-round; hours are from 11:30 a.m. until 9:00 p.m., seven days a week.

Hindman's Harborside Cash Market is a small store in the village of Swans Island. Steamed hot dogs and submarine sandwiches are available along with grocery items. Hindman is a photographer and stocks a good supply of film. Beer and liquor are not sold on this dry island. Store hours: summer 9:00 a.m. to 8:00 p.m.; winter 10:00 a.m. to 6:00 p.m. (207) 526-4130.

The General Store at Minturn has a lunch counter open from 11:00 a.m. to 2:00 p.m. and 4:00 p.m. to 8:00 p.m. Tuesday through Saturday. Monday hours are 11:00 a.m. to 2:00 p.m. This store stocks meats, fresh vegetables, fruits, and groceries, as well as drugstore items.

The Odd Fellows put on Sunday breakfasts during the summer; all you can eat for a reasonable fee.

What to Do

The Library / Museum next to the ferry landing has an interesting collection of Indian arrowheads, old tools, spinning wheels, an old-fashioned sled, photographs, and books. One room is furnished as an early American bedroom, with a hooked rug, old-fashioned bed with quilted spread, and a homemade cradle in the corner. Outside the museum is the bell used for many years at the lighthouse on Hockamock Head. Museum hours vary depending on availability of volunteers. It is usually open Saturday afternoons, year-round. Donations are appreciated.

Bicycling is popular on this island with its 27 miles of roads, of which 12 miles are paved. The numerous deep coves and inlets make for picturesque views.

The Carrying Place is a narrow stretch of land between Toothacher Cove and Back Cove. Wild roses grow along the sand here, and the views are extensive both ways. A good picnic spot is the pebble beach on the Toothacher Cove side of the Carrying Place.

Fine Sand Beach has unusually beautiful sand for the Maine islands. The beach is located on the west side of Toothacher Cove, accessible by a half-mile footpath running through the woods from the West Point road.

Hockamock Head has a picturesque unoccupied lighthouse marking the entrance to Burnt Coat Harbor. From this headland a visitor has a sweeping view of Harbor Island, Gooseberry Island, Marshall Island, and the open sea. The lighthouse was built in the 1800s and automated in 1974.

The lobsterman's wharves at Swans Island village on the west side of Burnt Coat Harbor are colorful spots. Both Fisherman's Co-op and Kent's Wharf handle buying and selling of lobsters.

Burnt Coat Gifts, in Nadia Sprague's home across the street from Hindman's Harborside Cash Market, offers handmade crocheted items, baby quilts, wool hats and other gifts.

Ocean Notions, along the road a bit north of the village, offers T-shirts, sweatshirts, handcrafts, quilts, and notepaper. The shop is open summers, Monday through Friday, 10:00 a.m. until 5:00 p.m. Off-season (June, and September through Christmas) hours are from noon until 4:00 p.m.

Four churches offer services: the Baptist Church in Atlantic, the Methodist Church and Church of God in Swans Island, and the Advent Christian Church in Minturn.

There is freshwater swimming in the granite quarry at Minturn, where quarried rock stepping-stones take the place of a ladder. Granite was quarried on Swans Island until 1925. The quarry is now filled with spring water that is a comfortable temperature for swimming—much warmer than the ocean.

A hiking trail leads from Goose Pond up Big Mountain, which offers a good view. The path is not well marked, however, and may be hard to find without local help.

The island has no public campground. Campfires are forbidden except on the beach well below high-tide mark. Please do not litter. Most of the wooded shoreland that appears deserted is privately owned, and visitors are requested to treat this property with respect.

School teacher Marie La Rosee with Frenchboro children.

Long Island

IN THE STILL of the morning, Frenchboro harbor is a beautiful sight. Each wharf, each lobster boat, the school, and the church tower are all perfectly reflected in the water. Walking along the shore on an early September day, I spotted 20 deer eating apples off the lower branches of the trees near Rebecca Lunt's home. Rebecca called them by name as they ate scraps of food from her hand.

Frenchboro is the village on Long Island. Situated seven miles southeast of Mount Desert Island and three miles east of Swans Island, Long Island is approximately 2½ miles in diameter. With the exception of the harbor area, the island's 2,500 acres are thickly wooded with spruce and fir. Moderately hilly, the island reaches its highest point at 210 feet above sea level.

The land on the west side of the main harbor rises gradually to a high ridge, with homes dotting the hillside. The entire population of Frenchboro (about 60 people in 1983) resides close to the long, narrow harbor. One mile of public road outlines the harbor, forming a horseshoe from the ferry landing on the east side of the harbor around to the last home on the west side.

Most of the island is forest wilderness once owned by David Rockefeller. With the exception of Rich's Head, Rockefeller passed the ownership of the land to his daughter, Margaret Delaney. The area has no homes or developed roads. Very little Long Island real estate comes up for sale. Since local residents own what little land remains, Frenchboro is one of the few Maine island communities that has not been affected by summer people.

Frenchboro is also off the beaten path for tourists. There is only one ferry trip on any given day so it is impossible to visit Frenchboro on a morning ferry and return on a later trip. One can come out on the ferry and return via the mailboat to Swans Island, taking the Swans Island ferry back to Bass Harbor, but the route is long and circuitous for only two hours on the island.

Since Frenchboro has no inn or licensed bed and breakfast, most visitors come by their own yachts and stay on a mooring overnight. Frenchboro residents don't miss the crowds. One of them commented about Acadia National Park, located on Mount Desert Island, "That's a good place for it. I'm glad it's not here."

One day each August this quiet little community is inundated with visitors. The twenty-first annual church fair, in 1983, drew 450 people, who came by lobster boat, motorboat, and ferry (which made two trips that day). Island women served a church dinner and sold homemade quilts, sweaters, afghans, Christmas decorations, and other items to benefit the church and the historical society.

Until 1982 Frenchboro had no telephone service to the mainland. The islanders had been requesting service for 12 years prior to installation. "We really appreciate it," says Vivian Lunt, "It's a great blessing." Vivian does the office work for Lunt and Lunt, the lobster dealer on the island, and she used to rely on the mail for all her business transactions.

The island has had electric power by cable from Swans Island since the 1950s. Individual wells provide the water. The well-drilling rig had been working on Frenchboro when I was there. I watched the operator maneuver his enormous rig onto the ferry. Although five of the six wells he dug came up with water, one drilling job left a family with a 450-foot dry hole.

Lunt is the most common name on Frenchboro. On my visit in 1983, I met ten of them — only about half of the Lunts on this island.

Danny Lunt took me out in his lobster boat while he pulled his wife's 15 traps. This was the "washing machine" run. His wife, Tina, had earmarked the money from her own traps to go toward the purchase of a badly needed washing machine. Tina and one-month-old Zachary were aboard for this part of Danny's day. He had already finished tending his own daily run of 300 traps. Using the electric winch, Danny pulled each trap and tossed the "keepers" into the bucket.

Tina Marie, Danny Lunt's lobster boat, is 31 feet long, with a cabin and deckhouse for protection in bad weather.

See Map On Page 102

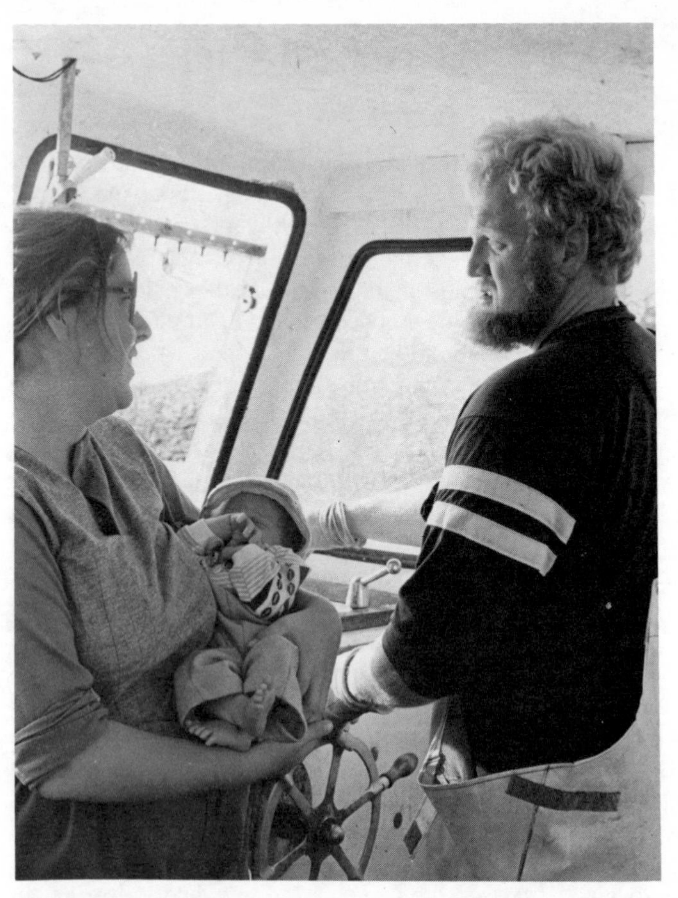

Frenchboro children experience lobstering at an early age. Danny, Tina, and Zachary Lunt on the Tina Marie.

Danny works alone during lobstering season, which slows down around Christmas. He then goes scalloping with his brother, David, in the winter.

History

Long Island was part of the Burnt Coat Group, a number of islands purchased in 1786 by Colonel Swan (see Swans Island history). Swan was an entrepreneur who bought, sold and mortgaged land with amazing rapidity. Long Island was traded back and forth: Swan sold it—in a group of islands—to his own agent, William Prince, in 1790; six years later Swan bought the group back.

By 1812 Swan's fancy footwork had landed him in a Paris prison on account of a debt he claimed he did not owe. From behind prison walls he continued his deals, mortgaging Long Island (and a group of islands) to Michael O'Maley, who took possession when Swan couldn't make the payments. The new owner offered to sell land to the settlers on Long Island. Between 1823 and 1835 Long Island residents gradually purchased the land on which they lived.

A community was well established by 1822 when Israel Lunt—the first of many Lunts on Long Island—built his store. There he fitted out vessels for the coastwise trade and was successful enough to be able to purchase about half the island in 1835. Lunt paid $600 for 1132 acres of land. According to Vivian Lunt's *Long Island Plantation: History of Frenchboro*, this appears to be the same tract of land recently owned by David Rockefeller.

Long Island and nearby isles were organized as Long Island Plantation in 1840. In 1850 a prominent lawyer, E. Webster French, offered to pull the political strings for a post office if the islanders would name part of the island for him. Thus the village — and post office — became Frenchboro.

The population grew from 19 in 1820 to its peak of 197 in 1910. A new school built in 1907 accommodated 60 pupils in two rooms. That building was still in use — as a one-room school with eight pupils — in 1983-84.

Frenchboro's school made the national news in the 1960s. Only two children attended in 1964, and the State of Maine threatened to cut off aid, thus effectively closing the school. Ingenious islanders invited 14 foster children to live in Frenchboro homes. Many of these children remained on the island until they went to high school off the island. At least two have returned to live there as adults.

Tony Brown was four when he came to live with Sanford and Vivian Lunt. Ten years later, the youth had to move off island for high school. In 1983, at the age of 22, Brown returned to Frenchboro, planning to go lobstering and scalloping for a living. "I like it; that's why I came back," says Brown. "This is my home; these [the Lunts] are what I consider my real parents."

How to Get There

The Maine State Ferry Service runs a single trip to Frenchboro, two days a week, year-round. The *Everett Libby* leaves from Bass Harbor on Mount Desert Island at 9:00 a.m. and returns from Frenchboro at 10:00 a.m. The crossing time is 50 minutes for the eight-mile trip. Mid-June to mid-

Unloading herring to be used for bait by Frenchboro lobstermen.

September the boat is scheduled on Wednesdays and Thursdays; during the rest of the year it goes on Thursdays and Fridays. For more information contact Maine State Ferry Service, Bass Harbor, ME 04653, (207) 244-3254.

The mailboat makes a daily round trip from Swans Island to Frenchboro and return. The boat leaves from Kents Wharf at the village of Swans Island about 12:30 p.m., arriving at Frenchboro about 1:00 p.m. The return trip leaves Frenchboro about 1:15 p.m. to return to Swans Island. (Check at the post office for accurate information on times of departure).

A short day visit to Long Island is possible, but circuitous. One can take the 9:00 a.m. ferry from Bass Harbor to Frenchboro, arriving at 10:00 a.m. After exploring Long Island for a couple of hours, return home via the mailboat to the village of Swans Island on Burnt Coat Harbor. One must hitch a ride across Swans Island to the Maine State Ferry Service terminal at Atlantic to catch the 3:00 p.m. ferry arriving at Bass Harbor at 3:45 p.m.

Lodging

There is no inn or licensed bed and breakfast on Frenchboro. Occasionally islanders will take visitors overnight. Make inquiries by writing Selectmen, Frenchboro, Bass Harbor, ME 04653.

Facilities for Yachtsmen

Gasoline, diesel fuel, oil, and water may be purchased at the Lunt and Lunt pier, the first wharf on the right when entering the harbor. The float has about three feet of water at dead low tide. Two or three moorings are available for yachtsmen for a small fee. Inquire at the Lunt and Lunt pier. The harbor is protected except from the northeast winds.

Meals

There are no restaurants on the island. As of 1983, the grocery store was closed. Inquire from local residents about whether it has reopened.

What to Do

Frenchboro Historical Society, located in the basement of Rebecca Lunt's house, has an interesting assortment of historical artifacts and books. The historical society sells island-made crafts in order to raise money, and the organization hopes to construct a permanent building in 1984.

A walk through the woods to the pebble beaches on the outer shores of the island is well worth the effort. Eastern Beach has smooth stones rounded by years of pounding surf.

In the Delaney / Rockefeller wilderness are two fire trails large enough for vehicles and several cleared paths leading from Frenchboro village across to Eastern Cove, Western Cove, and Deep Cove. Although there is no sand beach, Eastern Cove and Deep Cove have fine gravel beaches. Visitors are requested to respect the property, to avoid leaving any litter, and to be extremely careful with cigarettes and matches. Camping and fires are not allowed on this island.

Children enjoy the Pool at Great Cranberry.

The Cranberry Isles

NESTLED CLOSE TO THE southeastern shore of Mount Desert Island lie the five Cranberry Isles: Great Cranberry, Little Cranberry, Baker Island, Sutton, and Bear. They form the Eastern Way and the Western Way entrances into several of the harbors on Mount Desert.

Only Great Cranberry and Little Cranberry have year-round communities. Roughly half of Baker Island is in Acadia National Park land while most of Sutton belongs to vacation home-owners. Just off the entrance to Northeast Harbor is tiny Bear Island, which has an abandoned, although picturesque, lighthouse.

The passenger boats to the Cranberry Islands leave from Northeast Harbor, one of the busiest yachting centers along the Maine coast. Graceful, well-kept yawls and sloops are fastened to their moorings like thoroughbred horses awaiting the arrival of wealthy owners. Smaller sailboats moored beside them look miniature in comparison. Rowboats and dinghies transport sailors and guests back and forth from yacht to dock while children in outboard boats circle the fleet.

The *Sea Queen* is small compared to the car ferries of the Maine State Ferry Service. Passengers sit on rows of seats, inside the cabin in the stern, as the boat exits through the congestion into the open waters toward the Western Way. From there, passengers have an excellent view of Mount Desert Island's Cadillac Mountain (1532 ft.), the tallest peak on the New England coast, and the humped ridge of nearby mountains.

Great Cranberry

The boat docks at the north tip of Great Cranberry, known locally as Big Island or Cranberry. It was named for its cranberry bogs, but when the land was drained for mosquito control about fifty years ago, the cranberry plants suffered. Ferry captain Wilfred Bunker comments, "The cranberries left but the mosquitos didn't." Cranberries can still be found but they are not so plentiful. Great and Little Cranberry are low to the water and both have peat bogs known as heaths (locally pronounced "hayth").

Cranberry is the largest of the five islands. Measuring 490 acres, it's about two miles long by a mile wide. Approximately 75 year-round residents live on this island, with about 300 to 400 summer residents during 1983. One group of selectmen and one school board serve both Great Cranberry and Little Cranberry, although each island has its own one-room school for kindergarten through eighth grade.

Lining the beach near the ferry dock are several businesses, including Spurling Cove grocery store, the Islander take-out lunch window, The Whale's Rib gift shop, an artist's studio, and a boatyard.

Cranberry Isles

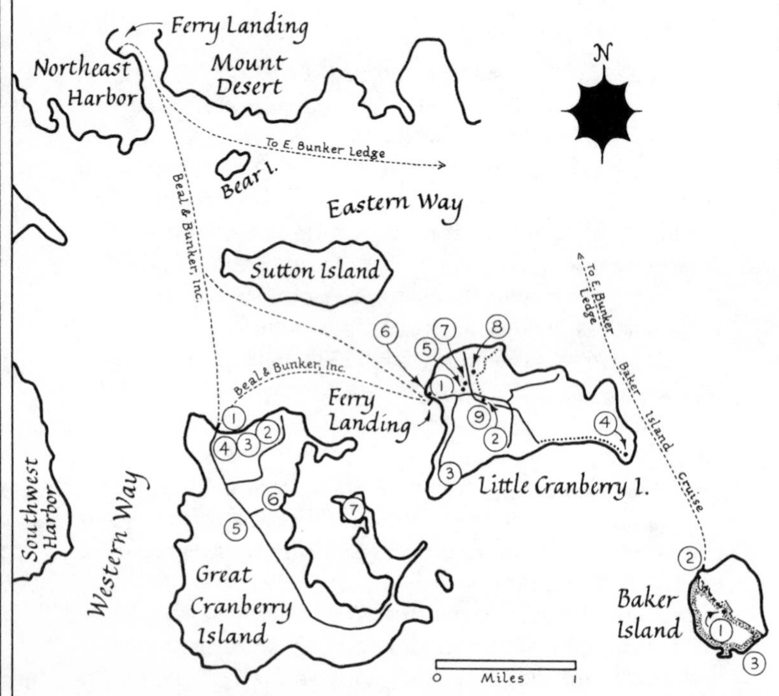

Islesford Village and Little Cranberry Island

1. Islesford Museum
2. Post office & Pine Tree Market
3. Graves of first settlers, 1769
4. Old Coast Guard Station (private)
5. Church
6. Restaurant
7. Neighborhood House
8. Heirloom Weavers
9. Island Bed & Breakfast

Great Cranberry Island

1. Ferry landing
2. Gift shop
3. Islander Take-out
4. Store
5. School
6. Boat yard
7. Fish Point

Baker Island

1. Lighthouse
2. Landing Beach
3. Dance Floor

Acadia Nat. Park

I rode my bicycle down the paved road that stretches the length of the island, passing houses, tennis courts, the fire station, and the schoolhouse, which doubles as the library. The road ends at a narrow isthmus and from there a path leads out on a hook of land to Fish Point. This is where islanders used to dry, salt, and smoke their fish.

A lobsterman's wife who was born on this island and has lived here many years says that about three-quarters of the houses are vacation homes. She says it was hard for her son to find a house to buy or even land on which to build. "The summer people are driving us out, slowly but surely," she comments. "When the older people pass away, it will be like Sutton [Island]." Sutton Island is predominantly owned by summer residents and has no year-round community.

Her husband notes, "The most hopeful sign is the new crop of youngsters on the island. I feel that it's the school what holds the island together." Although the one-room school had only six children in 1983, there were 11 preschool children on the island.

Tourism has increased in recent years. Some tourists, expecting to find stereotypes of the colorful Maine lobstermen and their families, stare at island residents or photograph them. One woman's annoyance was couched in quiet amusement as she related an incident: "I heard a little boy say to his mother, 'The people on this island don't look any different than they do anywhere else!'"

A saltwater cove known as "the pool" lies between Fish Point and the rest of the island. At the boatyard there, I met a family that had recently made their permanent home on Great Cranberry. The parents were relaxing on the dock, enjoying the sun and watching their young children play in the water. These former New Yorkers expressed surprise at how quickly islanders included them in the social life of the island — the potluck dinners and the card parties. They appreciated the fact that their children were always welcome at such gatherings.

Little Cranberry

Little Cranberry is small, only about 350 acres. From one end to the other it is only 1½ miles, and about ¾-mile at its widest. The village on Little Cranberry is named Islesford. Between 90 and 100 people live here year-round as of 1983–84, while the summer population peaks at about 400. The island attracts about 20,000 day visitors a year, most of whom come to tour the museum run by Acadia National Park. Near the museum are public rest rooms, a facility as rare on Maine islands as the Atlantic puffins of Matinicus Rock.

This tourist traffic supplements the island's traditional lobstering and fishing economy. Islesford has its own fisherman's co-op and wharf where lobstermen bring their catch at the end of each working day. The co-op in turn sells the lobsters to dealers on the mainland and also to the tourists.

A surprising number of Islesford residents are young people who have moved onto this island. They mix well with the islanders whose families have been there for generations. Karen Fernald, one of the partners of the Island Bed and Breakfast, is the daughter of an Islesford lobsterman and sister of three more lobstermen on this island. Her family can trace its ancestry on this island to the late 1700s.

One young mother whose family goes back several

generations on Islesford remarked that she likes bringing up her daughter here because of the feeling of security. "There's no crime, no sense of danger," she states. "I want to know what is happening to her."

Baker Island

While it is possible to walk from Little Cranberry to Baker Island across the bar on an extreme low tide, this route is not recommended for visitors. Tourists should take the cruise boat from Northeast Harbor. The Islesford Ferry Company runs two trips per day to Baker Island. The narrated cruise takes about 4½ hours to go to Baker, tour the island with a guide, and return to Northeast Harbor. The boat stops along the way to visit the osprey nest on Sutton Island and the seals on East Bunker Ledge.

Baker Island has no harbor and no pier, so the cruise boat catches a mooring in deep water offshore. Passengers climb into Grand Banks fishing dories and the guides row the heavy dories to the beach. Our guide gave us a tour of the island's nature and history. A United States Coast Guard automated lighthouse stands dramatically on the highest point in the middle of the island. I particularly enjoyed sunbathing on the huge boulders and flat ledges, one of which is known as Dance Hall Rock (some local people use the term the Dance Floor or the Dancing Rocks).

A trip to Baker is enhanced by reading a short paperback that is available either from Acadia National Park or in local bookstores: *John Gilley of Baker's Island*, by Charles W. Eliot, President of Harvard University for 40 years. Eliot was a

Beachcombing is a popular pastime along Great Cranberry's pebble beaches.

prominent founder of the Northeast Harbor summer colony.

The book covers the period from 1812 to 1896 on the Cranberry Isles, focusing first on William and Hannah Gilley, who raised a family on Baker Island, and then on their son, John Gilley. Baker Island, less than a mile across and with no good harbor for a boat, was home for the pioneering Gilleys and their 12 children. The family kept a yoke of oxen, six cows, a few cattle and hogs, and 50 sheep. The few items the Gilleys needed to buy were purchased with cash from the sale of butter, eggs, smoked herring, and sea-bird feathers. Each Sunday, weather permitting, Hannah and her children rowed seven miles to church in Southwest Harbor on Mount Desert Island.

John Gilley's life spanned from the pioneer days of the 1820s through the 1880s, when the first summer people began to come to the Cranberry Islands.

History

Pioneer settlers such as the Gilleys were not the first residents of the Cranberry Isles. Indian skeletons and utensils have been found buried near the boat houses on the sand beach at Little Cranberry.

Samuel de Champlain made the first recorded visit in 1604, but these islands were not settled by permanent white residents until after the French and Indian Wars. The first settler on Little Cranberry was Job Stanwood, in 1760, but he only stayed two years. The first permanent settler on the Cranberry Isles was Ben Bunker, in 1762. John Stanley came in 1769, followed by Sam Hadlock and Benjamin Spurling. The family names of Bunker, Stanley, Hadlock, and Spurling are all still common.

An excellent thumbnail sketch of the history of the Cranberry Isles is *The Town of Five Islands* by Ted Spurling of Islesford. Ted Spurling speaks with some authority; he has Stanwood, Hadlock, Spurling, Bunker, Stanley, and Gilley ancestors — all from the earliest settlers on the Cranberry Isles.

One of his most colorful ancestors was Benjamin Spurling, who was captured by the British navy during the War of 1812. Spurling had hidden his two ships in Norwood's Cove on Mount Desert Island because he feared they would be commandeered by the British. When word of this reached the British commander, he sent the man o' war *Tenedos* to capture the ships.

When Spurling refused to tell the British where his ships were kept, the British took him as a hostage aboard a rowing barge and headed for Norwood's Cove. As the story goes, Spurling's five sons called out the neighbors, and soon the cove's woods were filled with armed men. When the rowing barge was deep in the cove, the hidden soldiers opened fire on the British, who soon after freed Benjamin Spurling and departed, abandoning their plans to capture the ships.

How to Get There

Beal and Bunker, Inc. schedules five trips a day to the Cranberry Isles, making a circuit of Islesford (Little Cranberry) and Great Cranberry, stopping with the mail or upon request at Sutton. The *Sea Queen*, a 42-foot, 68-

passenger launch, leaves from Northeast Harbor on Mount Desert Island.

It's a short run to either island, taking 15 to 20 minutes to cover the two to three miles from Northeast Harbor. The trip between the two islands takes another 15 minutes. The earliest morning boat leaves Northeast Harbor at 10:00 a.m., and the last boat of the day leaves the Cranberry Isles about 8:00 p.m., so it is possible to go out for the whole day or just for a meal.

Beal and Bunker, Inc. also runs Islesford Historical Cruises aboard the *Sea Princess*, which leaves Northeast Harbor to visit the Islesford Historical Museum on Little Cranberry twice a day during July and August and once a day in June and September. The evening cruise has a dinner option (available at the Islesford Dock) before returning to Northeast Harbor. For more information, contact Beal and Bunker, Inc., Cranberry Isles, ME 04646, (207) 244-3575, open 7:00 a.m. to 7:00 p.m.

The Baker Island Cruise boat also leaves Northeast Harbor at 1:00 p.m. daily, mid-June to early September. Because cruises may be cancelled due to poor weather, travelers should check ahead with Islesford Ferry Co., Islesford, ME 04646, (207) 244-3366, open 9:00 a.m. to 5:30 p.m.

Facilities for Yachtsmen — Great Cranberry

Beal and Bunker dock has gasoline and diesel fuel and oil. The town dock has a float for dinghies.

Cranberry Island Boat Yard repairs and stores large motor or sail yachts. The owner is Ed Gray, Great Cranberry, ME 04646, (207) 244-7316.

Another boatyard, run by Charles Liebow, is situated near the stores on the beach near the ferry landing.

Meals — Great Cranberry

The Islander is a take-out window where customers may buy hamburgers, sandwiches, homemade soups, chowders, and ice cream. Seating is at picnic tables with umbrellas for shade. Summer hours are 8:00 a.m. until 9:00 p.m., from mid-June to the end of September.

Spurling Cove, the island store, is open year round and stocks a variety of grocery items. Hot coffee is available. Summer hours are 8 a.m. to 5:00 p.m., seven days a week.

What to Do — Great Cranberry

It is about a 2½-mile walk to Fish Point from the ferry landing. Bicycles can go as far as the paved section. Fish Point is private property but has no home on it. Please respect this property as well as all Cranberry Island land and do not leave litter.

Beachcombing is popular along the pebble beaches near the dock.

The library is open five days a week in summer, from 9:00 a.m. to 3:00 p.m.

The Whale's Rib gift shop features quilted jackets, canvas bags, cards, and gifts. Many items are handcrafted on the island. This shop is open seven days a week, 9:30 a.m. to 5:00 p.m. from the end of May until mid-September.

Islesford Dock restaurant and Cranberry Isles Fishermen's Co-op on Little Cranberry.

Lodging—Little Cranberry

Island Bed and Breakfast on Little Cranberry is situated in the center of town about ¼ mile from the ferry landing. This two-story frame house has several guest rooms including a two-room family suite. Spotlessly clean and tastefully decorated in early American style, this bed and breakfast offers the only lodging on any of the Cranberry Islands. It is open from mid-June to the end of September. For more information, contact Karen Fernald or Sue Jones, Islesford, ME 04646, (207) 244-9283.

Facilities For Yachtsmen—Little Cranberry

Little Cranberry has good facilities for yachtsmen. The Islesford Dock restaurant maintains four or five moorings available for overnight rental. Inquire from Conley Salyer, harbormaster and owner of the restaurant. People who wish to stop for lunch may tie up at one of Islesford Dock's three floats. There is about seven feet of water at low tide. Bags of cube ice are available at the restaurant. Salyer advises against anchoring in Hadlock Cove because of thick kelp on the bottom unless the yacht has a heavy fisherman-type anchor.

Islesford's yacht club schedules small sailboat races during the summer season. The clubhouse is a room on the same pier with the Islesford Dock.

Some of the graceful rowboats on this island were designed and built by Arthur ("Chummy") Spurling, an Islesford man who died in 1975 on his 102nd birthday.

Gasoline, diesel fuel, oil, and water may be purchased at the fishermen's co-op dock, the middle dock between the ferry landing and the restaurant.

Meals—Little Cranberry

The Islesford Dock restaurant is open for coffee and muffins at breakfast time. Lunch is served from 11:00 a.m. until 2:30 p.m., and dinner from 5:00 p.m. until 9:00 p.m. A separate take-out window offers lobster rolls, ice cream cones, and sandwiches, which may be eaten at the picnic tables on the deck. The restaurant is open from early June until late September. Call for dinner reservations during the busy summer season: Islesford Dock, Islesford, ME 04646, (207) 244-3177.

Pine Tree Market has an unusually good selection of food for an island store, including such items as fine wine, unusual cheeses, avocados, steak, and herb tea. The store is open year-round. Summer hours are 7:30 a.m. to 5:30 p.m. daily, closed Sundays.

What to Do—Little Cranberry

Exploring the roads and beaches of this small island, one might walk from the ferry landing down the paved road to the south shore beach. A United States Life Saving Station on this island's southeast point was manned by local islanders between 1879 and 1915. As one Islesford resident explained, "This was too good an idea for the government to keep." He said the men sent by the United States Coast Guard later could not compare with islanders in local knowledge. The Coast Guard abandoned this station about 1946 and the building is now privately owned.

Another nice walk is to Maypole Point on the southwest side of the island, only ¼ mile from the tip of Great Cranberry.

The Sea Queen *provides passenger service between Northeast Harbor and the Cranberry Isles.*

Hikers should remember that all beaches on Cranberry Isles are privately owned. They remain open to the public only because they are not abused or littered. Please cooperate by respecting these areas.

The Islesford Museum is located in a handsome brick building near the ferry landing on property owned and managed by Acadia National Park. The museum was built in 1927 at the behest of William Otis Sawtelle, a Haverford professor who summered on Little Cranberry. The interior brick walls and high, beamed ceiling set off an interesting collection that includes historical documents, books, scrolls, household goods, artifacts from shipwrecks, ship models, drawings, and old tools. A special exhibit shows how lobsters are caught. The museum features books and manuscripts of Rachael Field, of Sutton Island, who wrote about life on the Cranberry Isles in her books, *God's Pocket* and *Calico-Bush*. The museum is open from 10:00 a.m. until noon and 1:00 p.m. until 4:00 p.m. daily in July and August. In June and September, it is open on a limited basis. There is no admission charge.

Next to the museum, and also part of Acadia National Park, is the Blue Duck building, built in 1850 as a ship store and sail loft. There are public rest rooms in one end of this building.

Heirloom Weavers is a craft shop operated by Kathleen Bowman in her own home. The shop features handwoven mohair stoles, tweed ponchos, and many items made from hand-spun Maine island wool.

The gift shop in the back of the Islesford Dock features island handicrafts such as hand-knit sweaters, "ropework" artifacts and hand-braided rugs. The shop is open 10:30 a.m. to 3:00 p.m. from late June to late September.

Bird watchers may spot up to 100 species during a Maine Audubon Society fall weekend on Monhegan.

Special Programs and Vacations on Maine Islands

Audubon Ecology Camp

Audubon Ecology Camp in Maine, a field base of the National Audubon Society, offers two one-week workshops open to the public each August. Timed for the beginning of the bird migration period, these workshops give participants a chance to study both insect-eating birds and shore birds. It is not unusual for participants to spot as many as 130 species of birds over the course of the week.

The camp is situated on National Audubon Society's Hog Island, located in Muscongus Bay approximately 60 miles east of Portland and seven miles east of Damariscotta. This 330-acre wilderness island has varied bird habitats such as spruce / fir forests, salt marshes, and mud flats.

The workshops are open to anyone interested in learning more about birds. Participants choose their own level of study, ranging from beginner to advanced. Field trips include some offshore expeditions. Bird-banding sessions give participants opportunities to observe live birds close up. The daily program starts early (about 6:00 a.m.) and continues with field trips through the day. Evenings are for lectures or study. The camp has a library of books, records, filmstrips, and Maine bird study skins.

The workshops are limited to 55 participants. For an additional fee, the University of Maine offers one semester hour of credit for this workshop.

Write to Audubon Camps, 4150 Darley, Suite 5, Boulder, CO 80303, (303) 499-5409.

Hurricane Island Outward Bound School

Hurricane Island, a tiny island just to the west of Vinalhaven, once was home for a bustling granite industry. During the years 1870 to 1915, Hurricane Island granite was quarried and shipped to major cities for construction. Around 1878, a stable population of 600 was reported, acording to Charles B. McLane in *Islands of the Mid-Maine Coast*. Even in 1910 there were 256 residents. But when the granite quarry closed, the population moved off en masse. By 1926 the place was deserted.

Since 1948, this island has housed another bustling enterprise — the Hurricane Island Outward Bound School. The concept of Outward Bound was born during World War II as an attempt to teach young British sailors how to survive a disaster at sea. It was found that people exposed to a series of challenging situations developed strength of character, resourcefulness, and the ability to survive. Since that time the concept has been used by educators all over the world.

Hurricane Island Outward Bound School is one of several Outward Bound programs in the United States. The standard course here is the Maine Sea Program. Students sail among Maine islands in 30-foot, ketch-rigged pulling boats and learn navigation, sailing, and seamanship.

Rock climbing on the vertical cliffs of the old granite quarry is a part of the course. Each student also spends three days alone on one of the many tiny islands in Penobscot Bay. Equipment for this three-day solo includes a fish hook and line, a sleeping bag, a plastic sheet to use as a tent, a tin-can stove, a first aid kit, two quarts of water, a knife, and a journal.

This Hurricane Island Outward Bound solo might be the purest way to experience a Maine island—without distractions from other peoples' chatter, and with plenty of time to contemplate the world of nature, the sea, the fog, and the snails in the tide pools. In such an environment, one has a chance to get in touch with one's inner self.

Besides the standard courses, special programs are offered for women, for educators, for management training, and for "youth-at-risk." A natural history program focuses on wildlife and the ecosystem.

Visiting yachtsmen are welcome to come ashore to tour the island with a guide from the program.

For information, contact Hurricane Island Outward Bound School, P.O. Box 429, Rockland, ME 04841, toll free (800) 341-1744.

The Island Institute

The Island Institute is the resource management, research, and educational division of Hurricane Island Outward Bound School. Dedicated to the Maine islands as "a nationally significant resource," the organization publishes an *Island Journal* for exchange of information, organizes conferences about the future of Maine islands, and sponsors research, internships, and fellowships.

For more information, contact Philip Conkling, Executive Director, Islands Institute, Box 429, Rockland, ME 04841, (toll free) (800) 341-1744.

Damariscove Island Preserve

Damariscove is a 200-acre, rolling, treeless island about four miles southeast of Boothbay. Owned and preserved by The Nature Conservancy, this island is open to the public for day visits. Most visitors take their own boats. For those who have no boat, occasional field trips to this island are scheduled by The Nature Conservancy and by Maine Audubon Society.

The island is two miles long and a quarter mile wide, with a narrow isthmus dividing the northern and southern lobes. A long, narrow harbor at the southern end provides a landing area. Private boats may be pulled up on the shore at the head of the harbor. Do not land at the private wharf or walk on the property near the wharf because it is used and maintained by local fishermen.

Meadows of shrubs and grasses cover the island with bayberry, steeplebush, clover, *Rosa rugosa*, sedges, and many other plants. The northern section, Wood End, was once covered by a spruce forest that was destroyed by fire in the 1890s. This is now one of the major nesting areas in the eastern United States for common eider, as well as the nesting home for roughly 1,000 pairs of black-backed and herring gulls. Visitors are requested to refrain from exploring north of the isthmus during nesting season, April 1 through August 1.

This island is thought to have been settled in the early 1600s, as evidenced by the fact that Plymouth Colony obtained food from Damariscove after the hard winter of 1622. Sheep farming was popular here, dating back to the mid-1700s. A Coast Guard Life Saving Station built in 1896 was abandoned in 1959. In 1966 Mr. and Mrs. K. L. Parker donated the island to The Nature Conservancy.

During the summer months a caretaker on the island will provide information. Visitors are requested to observe these rules: no overnight camping, no fires, no pets ashore. Stay away from the East Tower (structurally unsound), Coast Guard Life Saving Station, and fishermen's cabins and wharf. Remove no plants, rocks, or animals.

For more information on Damariscove Island Preserve, contact Mr. and Mrs. Morley Roberts, East Boothbay, ME 04544, (207) 633-4781.

Damariscove Island Preserve is but one of many Maine saltwater preserves owned by The Nature Conservancy. These preserves make up the Rachael Carson Seacoast. Rachael Carson was a founder and honorary Chairman of the Maine Chapter of The Nature Conservancy. Only a few of these natural sanctuaries are open to the public. Visitors are requested to stay on established trails, to avoid littering, and to be sensitive to wildlife, seals, and sea birds. For brochures and maps of island preserves, contact Maine Chapter, The Nature Conservancy, 20 Federal Street, Brunswick, ME 04011, (207) 729-5181.

Maine Audubon Society

Maine Audubon Society schedules field trips to many island locations. Typical expeditions might be a bike trip to Chebeague Island or a visit to the home of Admiral Peary on Eagle Island in Casco Bay. A boat trip around Matinicus Rock allows participants to see the Atlantic puffins and other sea birds. Weekend trips are scheduled on Monhegan and at Acadia National Park on Isle au Haut.

These trips are open to the public for reasonable fees. Contact Maine Audubon Society, Gilsland Farm, 118 U.S. Route 1, Falmouth, ME 04105, (207) 781-2330.

Beal Island

Beal Island is a 64-acre wooded island on a saltwater estuary off Georgetown, Maine, not far from Bath. The island is owned and operated by the Appalachian Mountain Club for the use of members only and their friends. For a modest fee, it is possible to join the club and camp on Beal Island.

The location is a canoe lover's dream. Beal Island is only a 20-minute paddle across Knubble Bay from the mainland. From this island, one can explore miles of inland saltwater estuaries and bays. Those who like whitewater canoeing can try the tidal rips.

Facilities on the island are minimal. The campground is a sunlit meadow amid a cedar grove. It is situated on a bluff overlooking the South Beach where cold-water swimming is possible. The only structure on the island is a privy.

AMC has a fleet of eight canoes available for rent at Knubble Bay Camp, the launching area in Georgetown. It is possible to arrange to stay overnight in the self-service AMC cabin at this camp before going to Beal Island.

For information about camping on Beal Island, contact registrars Bob and Carol Rehn, 75 Wilshire Park, Needham, MA 02192, (617) 449-0723.

Beal Island is part of the AMC's tidewater canoeing program, which offers excursions of up to three days or longer. Recently, the AMC has developed a 60-mile "Down East" Coastal Canoe Trail from Portland to Pemaquid Point. For information, contact Appalachian Mountain Club, Joy Street, Boston, MA 02108, (617) 523-0636.

Maine Windjammers

One of the most adventurous ways to explore islands in Penobscot Bay is aboard a windjammer on a week-long sailing vacation. These cruise boats anchor in a different harbor each night and often stop at Vinalhaven, North Haven, Isle au Haut, and Swans Island, plus many smaller islands not mentioned in this book. Home ports are Rockland, Rockport, Camden, and Belfast.

A number of old-time sailing vessels, originally used in the coastal cargo trade or for fishing, have been renovated to carry passengers. The *Stephen Taber*, built in 1871, claims to be "the oldest documented sailing vessel in continuous service in the United States." Other old vessels include the *Mattie* and the *Mercantile*. Some windjammers, such as the *Mary Day*, were built in recent years specifically for this purpose. There are 12 sailing vessels in the Maine Windjammer Association , plus a number of independents for a total of 16.

Windjammers vary in size. Large ones such as the 120-foot *Adventure* and the 112-foot *Roseway* (one of the Tall Ships during the Bicentennial) hold 37 passengers. Smaller vessels, such as the 64-foot *Isaac H. Evans*, or the 78-foot *Sylvina W. Beal*, carry between 18 and 22 guests. Accommodations on these vessels vary from small double staterooms to bunk in a common room. At least one windjammer, the *Mary Day*, offers a no-smoking cruise.

Many of these boats have no motor and rely entirely on wind power. Camaraderie builds up as passengers hoist sails, lift the anchor, navigate, and take the helm under the direction of the captain and mates. Evenings are spent exploring ashore or singing and socializing on deck. One night may be devoted to a lobster bake on a deserted island.

Rates per person for the week vary from boat to boat. For brochures, rates, and addresses for each of these privately owned vessels, write to the Maine Windjammer Association, Box 317P, Rockport, ME 04856, (207) 236-4867.

Gulls are among the species of birds studied at Audubon Ecology Camp on Hog Island. From the Isles of Shoals to the Cranberry Isles, these hardy birds greet visitors.

Bibliography

Beveridge, Norwood. *The North Island: Early Times to Yesterday.* North Haven, Maine: N.H. Bicentennial Committee, 1976.

Boehmer, Raquel. *A Foraging Vacation: Edibles From Maine's Sea and Shore.* Camden, Maine: Down East Books, 1982.

Caldwell, Bill. *Islands of Maine.* Portland, Maine: Guy Gannett Publishing Company, 1981.

Conkling, Philip W. *Islands In Time.* Camden, Maine: Down East Books, 1981.

Daniels, Mrs. E. A. *Facts and Fancies and Repetitions about Dark Harbor by One of the Very Oldest Cottagers.* Cambridge, Massachusetts: Cosmos Press, 1935.

Duncan, Roger F. and Ware, John P. *A Cruising Guide to the New England Coast.* New York: Dodd, Mead & Company, 1983.

Eliot, Charles W. *John Gilley of Baker's Island.* Acadia National Park: Eastern National Park and Monument Association, 1899.

Hopkins, William. *Better Than Dying.* North Haven, Maine: Hopkins Publications, 1983.

Jewett, Sarah Orne. *The Country of the Pointed Firs.* Garden City, New York: Doubleday & Company, 1956.

Long, Charles A. E. *Matinicus Isle: Its Story and Its People.* Lewiston, Maine: Lewiston Journal Print Shop, 1926.

Lunt, Vivian. *Frenchboro: Long Island Plantation: The First Hundred Years.* Penobscot, Maine: Downeast Graphics, 1980.

Lunt, Vivian. *Long Island Plantation: History of Frenchboro.* Privately printed, 1976.

McLane, Charles B. *Islands of the Mid-Maine Coast.* Woolwich, Maine: The Kennebec River Press, 1982.

Rutledge, Lyman V. *The Isles of Shoals in Lore and Legend.* Boston: The Star Island Corporation, 1971.

Rutledge, Lyman V. *Ten Miles Out: Guide Book to the Isles of Shoals,* 5th ed. Boston: Isles of Shoals Association, 1972.

Rich, Louise Dickinson. *The Coast of Maine.* New York: Thomas Y. Crowell Company, 1962.

Simpson, Dorothy. *The Maine Islands in Story and Legend.* Philadelphia and New York: J. B. Lippincott Company, 1960.

Smith, Harry. *Windjammers of the Maine Coast.* Camden, Maine: Down East Books, 1983.

Spurling, Ted. *The Town of the Five Islands.* Reading, Vermont: Van Hauten Graphics, 1979.

Thaxter, Celia. *Among the Isles of Shoals.* Boston: Houghton, Mifflin, 1873; reprinted Bowie, Maryland and Hampton, New Hampshire: Heritage Books, Inc. 1978.

Westbrook, Perry D. *Biography of an Island.* New York: Thomas Yoseloff, 1958.

INDEX

Rates

Prices are based on 1983 rates, in most cases. Check current rates before you go. Names, addresses, and telephone numbers for boats, inns, and restaurants are listed in each chapter.

Isles Of Shoals

Viking Sun	Round trip stopover: adults: $10; children 5–12: $8.
Shoals Marine Laboratory	Non-credit courses for the general public range from $245 to $365 for five days.

Casco Bay

Casco Bay Lines	Bailey Island Cruise: adults: $9.75; children 5–9: $6.25. Mailboat Cruise: adults: $7.85; children: $5.25. Round trip stopover at Peaks: adults: $1.80; children: $.90.
Buccaneer to House Island	Round trip, fort tour, and clambake: $18 to $25/person.
Kristi K. to Eagle Island	Round trip: adults: $10; seniors: $8; children: $5.
Bustin Island boat	Round trip stopover: $5.
Water Taxi from Cousins Is. to Great Chebeague	One way: $1.75; bicycle: $2.
Moonshell Inn on Peaks Is.	July and August, rooms: $25–$38/night; off season: $23–$32. Prices include breakfast.
Beach Avenue House Long Is.	Spring and fall weekends: $75 per unit; summers: $200 and $275/ unit for first week.
Chebeague Inn	July 4 to Labor Day: rooms: $50–$70/night; mid-May to July 4 and Labor Day to Columbus Day: 20% off.

Monhegan

Laura B.	Round trip: adults: $14; children under 12: half-price. Parking near dock: $2/day; $12/week
Balmy Days	Round trip: adults: $18; children under 10: $14. Parking at Boothbay behind firehouse: no charge/24 hours. Chimney Pier lot: $3–$5/day; Sample's lot: $3/day.
Phalarope	Charters: $250/day minimum; water taxi: Boothbay to Monhegan: $125; Port Clyde to Monhegan: $90.
Island Inn	$33.50–$43.50/person/day, includes breakfast and dinner.
Monhegan House	Double room: $32; single room: $21.

Trailing Yew	Adults: $30/person/night; children under 10: $10–$20. Prices include two meals.
The Hitchcock House	Efficiency room (cooking facilities): $35/day; double room: $25/day; single: $19.
Tribler Cottage	Housekeeping apartments, mid-June to mid-September: $325/week; spring and fall: $300/week; winter: $40/day or $280/week; room: $36–42.

Matinicus

Stonington Flying Service	Mail plane passengers: adults: $15; children: $7.50.
Albert Bunker's boat	Charter service one way from Rockland to Matinicus: $150. Charter from Matinicus around Matinicus Rock and return: $55; stop to go ashore: $75.
Katz's cottages	Weekly rate: $235; weekend rate: $155.

Vinalhaven

Maine State Ferry Service*	Mid-June to mid-September: round trip, adults: $3.50, round trip, child: $2.00, round trip, auto: $15.25
Tidewater Motel	June 1 to September 30: double: $38–$45; single: $23–$31; off season double: $26–$31; single: $18–$25.
Bridgeside Inn	Double room: $24; single room: $18.

North Haven

Maine State Ferry Service*	Mid-June to mid-September: round trip, adult: $4.50; round trip, child: $2.50; round trip, auto: $18.75.
Almon Ames's guest house	Room: $25; dinner: $15; breakfast: $5.
Bulli Ruffian	Double room: $50; single room: $25; dinner: $25.
Mullen Head Park campsite	Donation requested
Haskell Village Camps	$155–$175/week.
Mabelle Crockett's guest house	$25/person/night, with two meals

Islesboro

Maine State Ferry Service*	Mid-June to mid-September: round trip: adult: $3; round trip: child: $2; round trip: auto: $15.25.
Islesboro Inn	$44–$74/person/day includes breakfast, afternoon tea, and dinner.
Dark Harbor House	July and August: $22–$24 person/night includes breakfast.

Isle au Haut

Mailboat/passenger service	Round trip: adults: $8; children: half fare; babies-in-arms: no charge; special trip: $30 to $50.
Acadia National Park	Campground shelter: $4/night.

Swans Island

Maine State Ferry Service*	Mid-June to mid-September: round trip adults: $2.50; round trip children: $1.50; round trip auto: $10.50.
Buswell's guest house	$20/person/night.

Long Island

Maine State Ferry Service	Mid-June to mid-September: round trip adults: $7.50; round trip children: $4.50; round trip auto: $30.50.

The Cranberry Isles

Beal and Bunker	round trip, adults: $3; children under 12: $2.
Baker Island Cruise	Adults $8; children: $4.
Islesford Bed and Breakfast	Double: $40; single: $35; family: $50.

Special Programs And Vacations

Audubon Ecology Camp	$385/person/week.
Hurricane Island Outward Bound	Maine Sea Program sailing expeditions: 6 days: $600; 11 days: $750; 22 days: $990.
Maine Audubon Society Trips	Day trips: non-member: $10–$25; weekend on Monhegan: non-member: $155–$180.
Beal Island	Camping: member AMC: $2/day; non-member with member: $3/day. Canoe rental: $6/day. Membership AMC: $25/year.
Maine Windjammers	Average rates: $350 to $400/week.

*Rates of Maine State Ferry Service are lower off season.